THE UNKNOWN GALEN

BULLETIN OF THE INSTITUTE OF CLASSICAL STUDIES SUPPLEMENT 77

GENERAL EDITOR: GEOFFREY WAYWELL

THE UNKNOWN GALEN

EDITED BY VIVIAN NUTTON

INSTITUTE OF CLASSICAL STUDIES
SCHOOL OF ADVANCED STUDY
UNIVERSITY OF LONDON

2002

BICS SUPPLEMENT 77

ISBN 0 900587 88 1

First published in 2002 by the Institute of Classical Studies,
School of Advanced Study, University of London,
Senate House, Malet Street, London WC1E 7HU.

Designed and computer typeset at the Institute of Classical Studies

Printed by Remous Limited, Milborne Port, Sherborne, Dorset DT95EP

CONTENTS

PREFACE

The Unknown Galen: Galen beyond Kühn was the theme of the joint symposium held in 1999 by the Institute of Classical Studies and the Wellcome Institute for the History of Medicine. Although as one speaker pointed out, many of the large tracts within the twenty volumes of Kühn's 1821-1833 edition could equally well be called unknown, the aim of the conference was to bring to wider scholarly attention some of the treatises that were not included in that standard edition. The choice of specific topic was left to the speakers, but with the proviso that it should reflect the enormously wide range of material, in a variety of languages, that is now available.

It is still growing. Since 1945 a new fragment of Galen, and often a whole work, has been announced or published on average every two years, and there is every prospect that this will continue for a few years yet. It has added another 25 per cent to the already enormous mass of the Corpus Galenicum, and represents the largest accession of writings to any classical author since the Renaissance. A brief bibliography, listing these unfamiliar works, is given as an appendix to this volume.

It is, of course, true that most of these works are not preserved in their original Greek. But Greek fragments continue to appear from among Byzantine scholia and theological writings, and one cannot entirely exclude the possibility of a major work being rediscovered among the unloved manuscripts of Byzantine medicine. Some full-length treatises have survived largely or only in medieval Latin, mainly in the preternaturally accurate translations of Niccolò da Reggio (active in the first half of the fourteenth century). Some of these were printed in the 1490 Latin edition of the *Opera Omnia Galeni*, and continued to be reprinted in later sixteenth-century editions of Latin translations of Galen. Paradoxically, it was the demands of the medical humanists for a return to Galen's original Greek that caused these genuine works to disappear from scholarly view until the early years of the twentieth century.

But the great majority of these 'new' Galenic texts come via another route, through a variety of translations into oriental languages, Syriac, Arabic, Hebrew, and Armenian. Although the existence of these versions had long been known in the West, little attempt to exploit this knowledge was made until the end of the nineteenth century. The first treatise to be published from the Arabic was perhaps the most striking, the second half of Galen's huge manual of dissection, *Anatomical Procedures (De anatomicis administrationibus).* Gotthelf Bergsträsser's publications of Hunain ibn Ishaq's *Risala,* a missive outlining what Galenic translations had been made into Syriac and into Arabic by 870, and of his list of works left out of Galen's autobibliography *De libris propriis*, alerted scholars to the potential value of this oriental material. Since then, there has been a steady stream of editions and translations, including works on philosophy, ethics, physiology, and many other aspects of medicine, enriching our understanding of Galen considerably. Even where the work itself is familiar, these translations often allow access to a lost Greek manuscript tradition older and better than that in our surviving manuscripts. More can be expected, since many major collections in the Islamic world have not yet been adequately catalogued and access to some of those that have been is not easy. There is always the likelihood, too, of finding an unknown Galenic treatise even in the best regulated libraries. Ivan Garofalo has recently announced his discovery in the Princeton University Library of the second part of Galen's own *Synopsis* of his *Method of Healing (Methodus Medendi),* a work previously known only from a quotation in Oribasius and from Hunain's report. It was included among the so-called *Alexandrian Summaries* of Galen's canonical treatises, but avowedly belongs to a different type of summary by Galen himself. The ongoing work of Gerrit Bos and Mauro Zonta on medieval Hebrew manuscripts of medicine and philosophy can also be expected to uncover further fragments of Galen.

These potential riches offer a challenge to manuscript hunters, particularly in Oriental libraries. But this conference also challenged classicists to make use of this material, even if that meant overcoming scruples about using translations of translations. They should not fear, for the level of translation by most orientalists is commendably high, and, even when there are mistakes, they are often easily noticed by someone familiar with the second, as opposed to the ninth, century. Besides, co-operation with an orientalist or a medievalist can bring intellectual benefits to both parties.

This conference could not have been held without the financial support of the Wellcome Trust, the academic assistance of Bob Sharples, or the organising skills of Frieda Houser and Margaret Packer. I am also grateful to Geoffrey Waywell and Richard Simpson for accepting this volume as a Supplement to the *Bulletin of the Institute of Classical Studies* and for their patience with a slothful editor.

IN DEFENCE OF KÜHN

VIVIAN NUTTON

The standard edition of the works of Galen, *Claudii Galeni Opera Omnia*, edited by Carl Gottlob Kühn, was published by the Leipzig firm of Karl Knobloch between 1821 and 1833. It cost 110 Thaler or, interleaved, 130 Thaler the set – approximately £16–10s or £19–10s in contemporary English money.[1] Like the Austrian musicologist Ludwig Köchel, Kühn has been made immortal with a K., for that is how classical scholars and medical historians have learned to refer to Galen, even though many of them have then gone on to point out his considerable deficiencies. There can be few such familiar names or initials who have endured such universal condemnation, and for such good reason.

Yet Carl Gottlob Kühn displayed many talents in his long life (1754-1840). He had a very wide acquaintance with Greek – not every classicist today could refer as he did to Eustathius for proof that to call doctors *demiourgoi* was to put them on a par with cakemakers and pastrycooks, which was certainly not what the unfortunate Dr. Ehrhardt had in mind when he used the word.[2] Eminent classicists, like Friedrich Jacobs in Munich, were his friends; and the young Wilhelm Dindorf, 'eruditissimus et amicissimus', is said to have done the work for his 1828 edition of Aretaeus.[3] Nor should one expect the long-term co-editor of the *Leipziger literarische Zeitschrift* to have been unfamiliar with the philological labours of his fellow Leipzig professor, the great Johann Gottfried Hermann.[4] Kühn knew how an editor of a classical text should set about his task. He should collate earlier editions, and investigate the

1 The price list at the back of my copy of the 1828 edition of Aretaeus, Leipzig, K. Knobloch, gives 5 thaler a volume. The other figures are derived from L. Choulant, *Handbuch der Bücherfreunde für die ältere Medicin* (Leipzig 1841) 113, and converted at the exchange rate for the Saxon thaler given in *Murray's Handbook for Travellers in Southern Germany* (London 1843) 11, a reference I owe to Mr John Symons. The two volumes together of Kühn's *Opuscula* cost 4 thaler 22 groschen. The suggestion that the the higher price for 'Schreibpapier' indicates interleaving was made to me by Prof. K. D. Fischer. The cost today would be in the range of £1,700-1,900.

2 C. G. Kühn, *Opuscula academica medica et philologica*, 2 vols (Leipzig 1827-28) vol. 2, 373, citing Eustathius, *Comm. in Odyss.* T. 135.

3 A. C. P. Callisen, *Medicinisches Schriftsteller-Lexikon der jetzt lebenden Aerzte, Wundaerzte, Geburtshelfer, Apotheker und Naturforscher aller gebildeten Völker.* 33 volumes (Copenhagen and Altona, 1830-45), *Nachtrag,* vol. 29, 1841, 376, distinguishing this edition and the reprint of Sprengel's *Dioscorides* from Kühn's own editions of Galen and Hippocrates. His comment is unknown to later editors of Aretaeus. The description of Dindorf comes from the flyleaf of a presentation copy by Kühn himself of his *Opuscula*, now in the Wellcome Library, London. How much Dindorf read of the volumes is unclear: vol. 2 remains largely uncut, and none of the chapters dealing with ancient medicine has ever been opened.

4 According to Callisen, *Lexikon* (above, n. 3), he was involved with this journal from 1803 onwards, and had also briefly edited the *Leipziger Gelehrte Zeitung* in 1783-84.

manuscript collections of a range of libraries in order to discover early readings or even new texts, and there is some evidence that he himself attempted to do this for Galen. His preface to Volume I refers to seeking out Scaliger's notes at Wolfenbüttel, those of Cornarius at Jena, and another set in the Royal Library in Dresden.[5] His edition contains several *editiones principes*: the first printings of the *Compendium on Pulses*, from a Copenhagen manuscript; the *Commentary on Humours* and *On the dissection of muscles*, from manuscripts in Paris; and the ending of *Quod animi mores* from a Munich manuscript transcribed for him by Jacobs. He was aware of the weakness of the material at his disposal, remarking on the seriously corrupt text of Chartier's edition, published a century and a half earlier and itself no improvement over the 1538 Basle edition, and he descanted on the overwhelming necessity for a new editor to emend or change the vulgate readings in very many places.[6] Kühn's aspirations and his awareness of what required to be done cannot be faulted.

Alas, he did not live up to his own protestations. His text of Galen, so all are agreed, is filled with errors and misprints of all kinds. At times, it is hard to tell whether a correction has been made accidentally by the printer or deliberately by the editor.[7] Indeed, in some of Galen's treatises the truth is more often to be found in the renaissance Latin translation at the foot of the page than in Kühn's Greek. He left out some genuine material in Latin translation that Chartier had included, while his own publication of the Commentary on *Humours*, though he was not to know it, made accessible to the world merely a renaissance forgery, to the subsequent confusion of the unwary or the uninformed.[8] But, contrary to what scholars have assumed, most of the task of collation and editing was done for him by others, like Jacobs. Indeed, the lavish praise bestowed on his friend Gottfried Heinrich Schaefer (1764-1840), in part trained in medicine, sometime Extraordinarius of Greek, and from 1818 to 1833 the Leipzig University Librarian, suggests that he, far more than Kühn, deserves to be regarded as the main source of editorial decisions and emendations.[9] Even Kühn's most impressive achievement turns out on closer inspection to be that of others. The 676 page index that constitutes volume 20 and which still retains its value today was the work of an assistant, Friedrich Wilhem Assmann, who cut down considerably the index prepared for the 1552 Juntine edition of the Latin Galen by the Ferrarese professor Antonio Musa Brasavola and supplied new page references.[10] All this makes for a severe, some might say a damning, indictment, and I shall not attempt to argue against it.

But Carl Gottlob Kühn deserves rather more than to be dismissed as a botcher of Greek, a philological parasite, a monster of overambition, a *Brasavola dimidiatus*. We need not be

5 C. G. Kühn, ed., *Claudii Galeni Opera Omnia,* 20 vols in 22 (Leipzig 1821-33) I, ix-xii.

6 Multa etiam reliquit quae cum linguae graecae rationi manifesto repugnarent, etiam codicibus manuscriptis invitis, corrigenda erant: ibid., xiii.

7 Cf. the story that to keep the printer happy Kühn tore out a few passages from Chartier and told him to get on with the job, D. Ehlers, ed., *Hermann Diels, Hermann Usener, Eduard Zeller, Briefwechsel*, 2 vols (Berlin 1992) I, 44.

8 He left out, for instance, *De motibus obscuris* and a fragment from the commentary on *Airs Waters, Places.* K. Deichgräber, 'Hippokrates De humoribus in der Geschichte der griechischen Medizin', *AAWMainz* 14, 1972, pp. 42-44. The situation is complex, since the renaissance forger (G. B. Rasario ?) included genuine material preserved elsewhere, e.g. in Oribasius or Maimonides, and some genuine Greek fragments.

9 Ibid., xiii. Details of his career and interests (principally in Greek prose and lexicography, both topics dear to Kühn) are given in the biographical notice of him by R. Hoche in the *Allgemeine Deutsche Biographie*, 65 vols (Leipzig and Munich 1875-1912), vol. 30, 1890, 524-525. By the 1820s Schaefer had already begun his famous feud with Hermann.

10 Claudii Galeni Opera Omnia, XX, vii-ix.

constrained by the somewhat lukewarm praise in his obituary, so redolent of a parish magazine, for being an excellent colleague and a devoted father. There is more to be said in his favour than that, for it has been his misfortune to have fallen among philologists who know only his failings and among medical historians whose gaze is turned more often to the exciting changes taking place in contemporary Paris. What I shall try to do briefly here is offer two different ways of looking at what he was trying to achieve in 1821: first, against the background of his other activities, and, secondly, against developments going on within German academic medicine during his lifetime. Only by doing this, can we gain a proper perspective on his edition of Galen.

First and foremost, we should remember that Kühn was the leading professor of medicine at one of the most flourishing medical universities of Europe.[11] Born in 1754, he began his academic career in 1785 as Extraordinarius of medicine at Leipzig, moving upwards to become Ordinarius of Anatomy and Surgery in 1802, and in 1804 Professor of Therapy. He was largely responsible for the remodelling of anatomical teaching in Leipzig, and for the major reorganisation of the medical faculty in 1810 which led to new chairs in particular aspects of medicine and to greater specialisation. He served as Rector magnificus of the university at the time of its quatercentenary in 1809.

Kühn himself was no narrow specialist: his interests and teaching spanned the whole of medicine. Hence, although after the reforms of 1810 he first taught medicine, he moved in 1812 to become Ordinarius of Surgery before spending the last twenty years of his life (he died in 1840) as Ordinarius of Physiology and Pathology. He was, one might say, the Pooh Bah of the Leipzig Faculty.

He was also a prolific author and editor with a wide range of interests – his publications fill twenty cramped octavo pages in Callisen's contemporary *Medizinische Schriftsteller-Lexikon*.[12] Therein lies both his strength and his weakness. At the same time as he was editing Galen, he was also seeing through the press Hippocrates, 1825-27, and Aretaeus, 1828, and writing on topics as disparate as recent archaeological discoveries of ancient surgical instruments and the dangers of eating poisoned sausages or mouldy cheese.[13] These years saw other publications besides; his *Opuscula academica medica et philologica*, 1827-28, his important revision of Blankaart's *Lexikon medicum* (Leipzig 1832), to say nothing of his work with various journals or his teaching commitments.[14] It is scarcely conceivable that he could have done more than cast the occasional eye over the manuscripts and proofs that his publishers were constantly sending across his desk.

But what the list of his publications does show is, first, his abundant energy, which lasted almost until the end of his long life. His Greek editions do not begin until he is in sixties and

11 Biographers depend heavily on the lists of his writings in Callisen, *Lexikon* (above, n. 3), vol. 10, 1830, 431-42; 29, 373-80; and in the *Neuer Nekrolog der Deutschen* 18, 1840-42, 720-24. A brief summary of his career, with two portraits of him, is given by Cornelia Becker, *Ärzte der Leipziger Medizinischen Fakultät* (Leipzig 1995) 23-26 and cover.

12 A. C. P. Callisen, *Lexikon* (above, n. 3), vol. 10, 1830, 431-42; *Nachtrag*, vol. 29, 373-80.

13 *De instrumentis chirugicis veteribus cognitis et nuper effossis* (Leipzig 1823); repr. in *Opuscula*, II, 306-19; *Versuche und Beobachtungen über die Kleesaure, das Wurst und das Käsegift* (Leipzig 1824), in part repr. in *Opuscula*, II, 175-384. His work on poisoning was partly written with his son, Otto Bernhard, who also became a medical professor at Leipzig in 1829.

14 Prof. Fischer reminds me of the usefulness of his series of *Additamenta ad elenchum medicorum veterum* (Leipzig 1826-37), which extends the list in Fabricius's *Bibliotheca Graeca*.

continue into his eighties. More interestingly, and this may come as a surprise, his bibliography reveals that he was far from being a dyed-in-the-wool conservative, opposed to innovation at all costs. In many respects he was among the most up-to-date medical men in Europe. He could without difficulty discuss surgery with Larrey, advise on the best construction and organisation of hospitals for foundlings, and ponder the effects of the blast and noise of the new artillery on the hearing of those nearby.[15] He was among the earliest and most vigorous proponents of vaccination, and he introduced into Germany the latest ideas and discoveries of the Hunters, Thomas Beddoes, and Charles Bell, among others.[16] He was, in short, an extremely talented publicist, concerned to make available to a German or a latinate European audience quickly and relatively cheaply the most important new discoveries in medicine and surgery. He took pains to ensure that the message was transmitted in the most effective way. Take for example *Der Gesundheitsfreund,* of 1817, his translation of Richard Reece's *The medical guide for the use of the clergy, heads of families and practitioners of medicine,* chosen because of its immense practicality. Kühn appears to have carried out the translation himself, or rather the adaptation, for, as he remarked in the preface, a mere word for word translation would not help to plant its ideas on German soil.[17]

It is in this light that one should view Kühn's editorial labours: he sought to produce an accessible and convenient text, supplemented with whatever useful documentation might be to hand. His editions were thus, in more ways than one, a German precursor of the Abbé Migne's *Patrologiae* or of a modern reprint house. His *Aretaeus*, 1828, includes the commentaries of Petit, Wigan, and Triller, while his Galen begins, xvi-cclxv, with the very valuable summary of Galenic biography and editions by Johann Gottfried Ackermann, lifted from Harles' revision of Fabricius – and how many of us have ever consulted Fabricius' original ? His index is still the easiest and simplest way to find many Galenic references – less cumbersome, and occasionally fuller, than the new Greek printed index or Brasavola's old Latin, to say nothing of Ibycus. The actual format of Kühn's editions makes them relatively easy to use, except perhaps when one has to search through thirty or forty pages to find a single chapter number. The simple act of providing page references to Chartier and the Basle edition simplifies enormously the task of finding a line or a word in these daunting folios. Above all, merely by publication, he made the writings of Galen more easily accessible – and for that we continue to give thanks.

I have so far argued that we should see Kühn in the guise, not of a philological editor, but of a medical publicist, a reprint specialist. Similarly, we should view the firm of Knobloch as hard-headed businessmen exploiting the new technologies of printing to make money – after all, even today, university presses are rarely organs of charity, whatever authors might wish. But this argument in Kühn's defence depends on a further hypothesis, that what was being produced was likely to find a sympathetic audience, or, to put it another way, that the study of ancient medical texts was still seen as valuable, indeed essential, for contemporary medicine.

15 C. G. Kühn, *Opuscula*, I, 356-61 (originally 1814); I, pp. 269-90 (originally 1810); I, 368-75 (originally 1814).
16 C. G. Kühn, *Die Kuhpocken* (Leipzig 1801); *De morbo vaccino-varioloso, Opuscula,* I, 128-259 (originally 1801-26); for his translations and editions of foreign texts, see Callisen, *Lexikon* (above, n. 3) 10, 440-41; 29, 38-40.
17 R. Reece, *Der Gesundheitsfreund (*Leipzig 1817) V.

Here two recent studies, by Thomas Broman and Hans-Uwe Lammel, have thrown considerable light on what was for many years a dark corner of medical history.[18] They have shown that in Germany down to the 1830s, if not later in some universities, the dominant form of learned medicine was eclectic in its orientation, rejecting an adherence to any one theory in favour of a therapeutic synthesis that brought together successful, or potentially successful, treatments, from a variety of sources. Great emphasis was placed on the observation of both patient and disease, following the model of Hippocrates, the father of true medicine – note here Kühn's reedition of Thomas Sydenham, the English Hippocrates (Leipzig 1826), and his special interest in both Aretaeus, the Hippocratic nosographer, and Caelius Aurelianus, whose books on acute and chronic disease contain the most detailed descriptions of disease in Latin.[19] The true practice of medicine depended on the accumulation of various kinds of direct experience, which were then subjected to clinical judgment. Physiology and pathology were both extremely valuable, but only in as much as they acted as propaedeutic disciplines for the true bedside physician.[20] In all this, history, and particularly ancient history, had an important place. It revealed the method of healing laid down by Hippocrates, a synthetic vision that had little sympathy with sectarianism and that emphasised the healing power of Nature.[21] It thus encouraged both the deeper understanding of Nature in all its forms and a critical interpretative attitude towards ancient texts, which were to be reviewed and consulted with as much care and intellectual awareness as modern reports of therapies from Baden or Bordeaux. Based on this accurate understanding, the modern doctor could select the best therapies from a wide variety of sources without falling into the trap of believing in fashionable speculations like Brunonianism. This medical theory, with its division of diseases into sthenic and asthenic, had enjoyed a great vogue in Germany in the first decade of the nineteenth century, but was now on the wane, and its major intellectual centre, Bamberg, had been demoted to being a mere surgical academy.[22] Its theories were denounced by the eclectics who, at the same time, were willing to adopt some of its therapies provided that they saw a clear benefit in them.

The eclectic method was spread in books and journals by learned physicians in many universities, whose views of contemporary developments ranged from the hostile to the emollient. At Jena, Christian Gottfried Gruner (1744-1815) was assembling and reprinting a variety of ancient and medieval texts on diseases like the English Sweat or syphilis.[23] Galenists know Gruner for his fallible publication of Cornarius' annotations to Galen, but they may be less familiar with his translations of Hippocrates (especially the so-called

18 Thomas H. Broman, *The transformation of German academic medicine, 1750-1820* (Cambridge 1996): Hans-Uwe Lammel, *Zwischen Klio und Hippokrates. Zu den kulturellen Ursprüngen eines medizinhistorischen Interesses under der Ausprägung einer historischen Mentalität unter Ärzten zwischen 1750 und 1830 in Deutschland,* Habil. Diss., Rostock 1999.

19 *Opuscula*, II, 1-127, 150-90, largely reprinting the notes of D. W. Triller.

20 Broman, *Transformation* (above, n. 18) 142f.

21 Ibid., 140.

22 Ibid., 128-58; W. F. Bynum and R. Porter, eds, *Brunonianism in Britain and Europe, Medical History,* Supplement 8, 1988.

23 C. G. Gruner, *Scriptores de Sudore Anglico superstites* (Jena 1847); *De Morbo Gallico* (Jena 1793). E. Giese, B. von Hagen, *Geschichte der medizinischen Fakultät der Friedrich-Schiller Universität Jena* (Jena 1958) 316-25.

spurious texts) and of the fragments of Euryphon, Diocles, Praxagoras and Chrysippus.[24] Gruner was an out and out conservative, denouncing 'the ruling spirit of the age which is interested in idolatry, in enlightenment and superstition, in revolution and recasting of all things;.... which hates everything old and seeks through the new to dazzle and mislead.'[25] He could certainly have been included among those who rejected modern ideas on the grounds that when they disagreed with what had gone before, they were dangerous, and when they agreed, they were unnecessary. But his was an extreme position, and recognised as such by his contemporaries. Many others, like Ernst Gottfried Baldinger at Marburg, or the historically talented August Friedrich Hecker at Erfurt, were more sympathetic to new developments. Hecker's journal was called the *Journal der Erfindungen, Theorien und Widerspruche in der Natur- und Arzneiwissenschaft (The Journal of Discoveries, Theories, and Controversies in Science and Medicine)*: all parts of the title are equally significant.[26] The great historian of medicine from Halle, Kurt Polycarp Joachim Sprengel, was another example of this movement. His *Versuch einer pragmatischen Geschichte der Medicin* (*An essay on a pragmatic history of medicine)*, first published in 1792 and revised in 1800 and 1821, sought to pick out the stages in the history of medicine that were most relevant to current concerns.[27] He himself was a leading botanist, and wrote a major textbook on medical semiotics.[28] There were many others: Karl Joseph Hieronymus Windischmann in Bonn, whose *Versuch über den Gang der Bildung in der heilenden Kunst* (*An essay on the place of cultivated Learning in the healing art)* appeared in 1809;[29] Eduard Löbenstein-Löbel in Jena, who used the evidence of the classical surgeons Herodotus and Antyllus to promote the medical virtues of sunbathing;[30] or Karl Friedrich Heinrich Marx, whose 1824 Göttingen dissertation *Origines contagii* linked ancient discussions of contagion with modern findings, and used classical evidence in order to give precision to modern terminology.[31] Above all, this approach to medicine was advocated by the most influential physician of the day, Christoph Wilhelm Hufeland (1762-1836).[32] In his *Journal der praktischen Arzneikunde und Wundarzneikunst (The Journal of practical medicine and surgery*), the leading medical journal of the time, in his teaching at Jena and then in Berlin, and in his contacts with the Prussian government, Hufeland propagated an eclectic medicine that sought to reconcile traditional, classical medical learning with new discoveries and theories. His aim was to produce not just good anatomists or researchers, but good physicians. He expected all

24 C. G. Gruner, *Jani Cornarii Conjecturae et Emendationes* (Jena 1789); *Bibliothek der alten Aerzte* (Leipzig 1780-82).

25 Quotation modified from Broman, *Transformation* (above, n. 18) 84.

26 Ibid., 85. Hecker was the father of the Berlin Professor and historian, J. F. C. Hecker, who continued the eclectic approach into the 1850s. According to Callisen, Kühn was a long-time editor of the *Magazin der Erfindungen,* but I have been unable to check whether this was the same as Hecker's journal.

27 K. P. J. Sprengel, *Versuch einer pragmatischen Geschichte der Medicin* (Halle 1792); 2nd ed., 1800; 3rd. ed, 1821. Lammel, *Klio* (above, n. 18) 221-41, 280-86.

28 K. P. J. Sprengel, *Handbuch der Semiotik* (Halle 1801).

29 Note also his *Ueber Etwas, das der Heilkunst Noth thut. Ein Versuch zur vereinigung dieser Kunst mit der christlichen Philosophie* Leipzig 1824).

30 E. Löbenstein-Löbel, 'Wichtige Ansichten über die Realisierung der Idee eines Sonnenbades', *Journal der praktischen Heilkunde* 6 (1815) 56-85.

31 K. F. H. Marx, *Origines Contagii* (Karlsruhe and Baden 1824); *Additamenta ad Origines Contagii* (Karlsruhe and Baden 1826).

32 Giese and von Hagen, *Geschichte* (above, n. 23) 375-88, describe his early career at Jena.

teachers of medicine to be truly Hippocratic, building solely on the basis of experience and a thorough study of classical literature, and free from any passion for systems or sectarian zealotry.[33]

This was the approach favoured by Kühn, an approach that looked back to history in seeking to pick out the most effective theories and practices from the past as a guide to the present. The key words that stud the books and articles of its protagonists – practical, pragmatic, learned, empirical, and the almost untranslatable *Bildung* – express clearly their attitude towards medicine as an art, a *Heilkunst*, of which the doctor was both priest and prophet. It was an approach that won many supporters, and that drew many students to its teachers. Jena and Leipzig in the first quarter of the nineteenth century were among the largest, if not the largest, medical schools in Germany.[34] Seen in contemporary perspective, medical eclecticism had many merits, not least its linkage of old and new, its willingness to consider together the best of the past and the best of the present.

It is in this context, far more than in that of classical philology, that Kühn's edition of Galen must be considered. A decade later, ideas on medicine had changed. The clinic and, gradually, the laboratory replaced the library as the workplace of the medical professor. The ancient medical writers became the object of historical study, not primary sources for modern therapeutics.[35] Kühn's Galen could now be bought for half its original price, a mere 44 Thaler (88 for the interleaved edition).[36] It stood undefended as a classical text, whose deficiencies have rightly aroused the scorn and derision of all subsequent Galenists. But it was not with philologists primarily in mind that Kühn had embarked on his edition, but his fellow students of the medical art. In doing so, he preserved and made accessible again the most important medical author of Antiquity for medical students who could read Greek, or at least Latin. At the same time, he gave subsequent editors of Galen a convenient base from which to begin. The fact that so many of the Galenic texts he did not publish remain almost unknown today to classicists and historians of medicine, is both a tribute to Kühn's success in cornering the field and a dreadful warning of the power of a standard edition to impose a straightjacket on subsequent generations. But that, one can cheerfully say, is not the fault of Carl Gottlob Kühn.

University College London

33 Broman, *Transformation* (above, n. 18) 184, citing Hufeland's 1813 recommendation to Berlin of the Breslau professor of medicine, K. A. W. Berends.
34 Broman, ibid., 28, gives some figures.
35 The change can be seen elsewhere, in the publications of the Sydenham Society in England, and even within the volumes of the great edition of Hippocrates by Littré.
36 Choulant, *Handbuch* (above, n. 1) 113. I am assuming no revaluation of the Saxon thaler between 1833 and 1841.

GALEN'S *ON MY OWN BOOKS*: NEW MATERIAL FROM MESHED, RIDA, TIBB. 5223

VÉRONIQUE BOUDON

Galen's *On my own Books*, one of the most frequently cited of all Galen's treatises, is at the same time one of those that has proved most frustrating to editors.[1] In his Leipzig edition of 1830 Carl Gottlob Kühn produced no major improvements to the Greek text, generally being content simply to reproduce the work of his predecessors.[2] Iwan von Müller, the most recent editor of this treatise, in 1891, had to admit his inability to present the reader with anything like a correctly established text.[3] He could do no more than demonstrate the dramatically lacunose condition of this small tract, and also of its companion, *On the Order of my own Books (De ordine librorum suorum)*. As he said, 'The text of these books has reached us today in a dreadfully damaged condition; whole words are missing, even whole sentences. Some parts of it have disappeared completely so that one might more appropriately call *On the Order of my own Books* and *On my own Books* fragments rather than real books.[4] In this paper I shall discuss the manuscript tradition of *On my own Books (De libris propriis)* as represented by the single, lacunose Greek manuscript, and examine the new evidence provided by the Arabic translation made by Ḥunain ibn Isḥāq, which is itself also preserved in a single manuscript.

The unique Greek manuscript is Milan, Ambrosianus graecus 659 (generally known as A). It has many large gaps in it, and served as the model for the *editio princeps*, the Aldine Galen of 1525. Later editors, down to Kühn and Müller, have largely reproduced its readings. There are also two Latin translations of the sixteenth century: the first, by Jean Fichard (Fichardus) appeared in 1531, the second by John Caius in 1556. These are, however, of only slight value, since Fichard based his translation on the Aldine edition, Caius directly on A.[5]

In these somewhat unfavourable circumstances, only the arrival of new material can allow us to make substantial progress. More specifically, it was absolutely essential to get hold of Ḥunain's Arabic version of *On my own Books*. This was far from easy, for the unique Arabic

1 I wish to thank Vivian Nutton for translating my text into English.

2 Galen, XIX 8-48 K.

3 I. Müller, *Claudii Galeni Pergameni Scripta Minora* (Leipzig 1891) vol. II, 91-124.

4 Ibid., LVI: Textus igitur libellorum ad aetatem nostram pessime mulcatus pervenit; cum permultis locis lectionis pravitate laborat, tum detruncatus miserrime est; desunt singulae voces, desunt sententiae integrae, quid ? quod partes aliquae deperditae sunt, ut fragmenta potius quam libros περὶ τῆς τάξεως τῶν ἰδίων βιβλίων et περὶ τῶν ἰδίων βιβλίων exstare dicas.

5 For the relationship between the Aldine Galen, the Latin translations and A, see V. Boudon, 'John Caius (1510-1573) traducteur de Galien: le cas du *De libris propriis* et du *De ordine librorum suorum*', in A. Garzya, J. Jouanna, eds, *I testi medici greci. Tradizione e Ecdotica* (Naples 1999) 17-27.

manuscript of this translation is preserved in a religious library in north-east Iran, at Meshed. Thanks to the help and determination of a French researcher working there, and through a complex system of exchange deals, I was recently able to obtain a set of photocopies of the manuscript, fortunately of a reasonably good quality.[6] This paper presents the first results of my reading of this manuscript.

First, however, one must recall what was said of our treatise by Ḥunain ibn Isḥāq in his famous *Risāla*, the letter in which he described at length the early Arabic and Syriac translations of Galen, since I want to compare what Ḥunain says with the treatise as we have it today. Ḥunain writes as follows: 'Galen's *Pinax*, in which he describes his own books, is composed of two parts: in the first Galen enumerates his works on medicine, in the second those of logic, philosophy, rhetoric, and grammar. In numerous greek manuscripts we have found the two parts put together as if to form as single whole. In them, he (i.e. Galen) planned to describe the treatises he had written, his intended aim, why he decided to write them, who he wrote them for, and when. Before me, Ayyub al-Ruhawi al-Abrash (the spotty) had translated them into Syriac; I translated them into Syriac for Dawud al-Mutattabib (the practitioner) and into Arabic for Ga'far Muhammad Ibn Musa.[7] However, since Galen never managed to produce a complete list of his writings, I added to the two parts of my Syriac translation a third, shorter section. In this I showed that Galen in this tract had failed to include a list of a portion of his books, and I enumerated many of them, in as much as I had seen or read them, and gave the reason why he had left them out.'[8]

That is what Ḥunain says. The first point to note here is that there is no trace of any division into two sections either in Kühn's edition, in the *editio princeps* or in the Milan manuscript on which both depend. But had this division existed in Ḥunain's Greek manuscripts, it would have come between chapters 1-10, which effectively deal with medicine and the different medical schools, and chapters 11-17. Nonetheless, we shall see in a moment that the Arabic translation from Meshed adopts a division into chapters that only partially corresponds to that of the Greek text.

More valuable perhaps is the information provided on the successive translators of this treatise, first Ayyub al-Ruhawi, whose Syriac version is now lost, and then Ḥunain himself, who tells us that he made two translations, the first into Syriac (the language of the Near Eastern Christians of the ninth century) for Dawud al-Mutatabbib; the second into Arabic for Abu Ga'far Muhammad ibn Musa. Ḥunain also took the trouble to provide a full list of Galen's books by adding a third section to his Syriac translation. This third section survives today in a later translation into Arabic, which one may consult in the edition and German translation of Gotthelf Bergsträsser.[9] Just as he had said in his *Risāla*, Ḥunain begins by

6 I am most grateful to my colleague Ziva Vesel who kindly obtained the manuscript for me in Meshed.

7 For details of these individuals, see F. Micheau, 'Mécènes et médecins à Bagdad au IIIᵉ/IXᵉ siècle', in D. Jacquart, ed., *Les voies de la science grecque. Études sur la transmission des textes de l'Antiquité au dix-neuvième siècle*, Hautes études médiévales et modernes 78 (Geneva 1997) 147-79. The first two Syriac translations are now lost.

8 G. Bergsträsser, ed., *Hunain ibn Ishaq, Über die syrischen und arabischen Galen-Übersetzungen, Abhandlungen für die Kunde des Morgenlandes* XVII.2 (Leipzig 1925), 3.

9 The Arabic is preserved in Istanbul, Aya Sofya 3590, fols 34r-37r (6/13th century), see G. Bergsträsser, ed., *Neue Materialen zu Hunain ibn Ishaq's Galen-Bibliographie*, Abhandlungen für die Kunde des Morgenlandes, XIX 2 (Leipzig 1932). See also M. Meyerhof, 'Über echte und unechte Schriften Galens nach arabischen Quellen', *SBAW* 1928, 534-41.

explaining why Galen left out some of his writings from his own catalogue: either Galen had not yet written them, or he no longer had copies and considered them lost, or they had been destroyed in the great fire of his library at the Temple of Peace and Galen had no plans for their rewriting. Ḥunain then provides a new list of Galen's genuine writings, before going on to list those that should be considered suppositious.

That, then, is Ḥunain's evidence for the text of this treatise as it was in the ninth century of our era, both as regards Greek manuscripts (and note that Ḥunain claims to have seen several) and the Syriac and Arabic versions.

The Greek tradition of Ambrosianus gr. 659 (Q 3 Sup)

If one now compares Ḥunain's description with the text printed by Kühn, and even with the considerable changes for the better introduced by Müller, one cannot help but notice several differences. But since all our editions go back to the Milan manuscript, Ambrosianus graecus 659 (Q 3 Sup.), we need to dwell for a moment on that. The Ambrosianus is a paper manuscript, 272 folios long, written in the fourteenth century, and containing fourteen works by Galen.[10] *On my own Books* occupies folios 187 recto to 197 recto. A careful examination of the quires that make up the manuscript reveals the loss of the outer bifolium of the 24th quaternion, that is at present folios 193-198. This material accident to our unique Greek manuscript has produced a major lacuna in both this treatise and *On the Order of my own Books,* which takes up the following leaves, folios 197 recto to 200 recto. This gap deprives us of some important comments on the anatomical writings of Marinus and Lycus of Macedon, for our text breaks off at the end of folio 192 verso, half way through chapter 3 (XIX 30,4 K. = 108,14 Müller). The missing text corresponds to sixty lines in A, or about 4 pages in Kühn - the Aldine editors of 1525 left 42 lines blank, but their hopes that it would one day be possible to fill in this gap in Greek have not so far been fulfilled.[11]

The Arabic tradition: Ḥunain's translation

In the absence of any surviving Syriac translation, we must turn to Ḥunain's Arabic version. The existence of the Meshed manuscript was first mentioned in 1970 by Fuat Sezgin in his bibliography of Arabic medical manuscripts.[12] Meshed, Rida, tibb. 5223, was said by Sezgin to contain Ḥunain's version on folios 22^{v} to 40^{v}, and to date from the seventh century AH (our thirteenth century). Sezgin went on to give a transcription of the titles of the various chapters. Alas, not everything he says is right, and his account needs correction in several places.

10 See A. Martini, D. Bassi, *Catalogus codicum graecorum bibliothecae Ambrosianae*, 2 vols (Milan 1906) II, 738-40. For a recent bibliography of this manuscript, see V. Nutton, ed., *Galen, On my own opinions* (Berlin 1999), 17; B. Mondrain, 'Jean Argyropoulos professeur à Constantinople et ses auditeurs médecins, d'Andronic Éparque à Démétrios Angelos', in C. Schulz, G. Makris, eds, *ΠΟΛΥΠΛΕΥΡΟΣ ΝΟΥΣ. Miscellanea für Peter Schreiner zu seinem 60. Geburtstag* (Leipzig 2000) 223-49.

11 P. N. Singer, in his English version of this treatise, *Galen. Selected Works* (Oxford 1997) 13, and note, p. 400, overestimates the size of lacuna as six pages, and is thus forced to speculate on its further possible contents.

12 F. Sezgin, *Geschichte des arabischen Schrifttums,* III (Leiden 1970), 78, no. 1. Its existence was apparently unknown to M. Steinschneider, *Die arabischen Übersetzungen aus dem griechischen* (Leipzig 1897) and M. Ullmann, *Die Medizin in Islam* (Leiden, Cologne 1970), the other standard repertories of information on such manuscripts.

Even before I had access to the Meshed manuscript, it was clear that Sezgin's ten chapters did not correspond to the order of chapters in the Greek; some were totally new, while others present in Greek were missing in Arabic. My reading of the manuscript confirmed at once that the Arabic offered in several places a fuller text than the Greek, and, conversely, that the order of the leaves had been seriously disturbed. By examining the manuscript and its quire numbers in detail, I have reached the following conclusions:

1. The 40 folios, plus one blank, are divided into ternions and quinions with 19 lines to the page. Script and layout suggest an earlier date than Sezgin proposed, probably the fifth century A.H. (our eleventh century), when there was still a substantial interest in Greek science compared with two centuries later.
2. Our manuscript has very strong links with another Meshed manuscript, Rida, tibb 5214/1, which contains *On the Order of my own Books* in Ḥunain's translation. The format, script, layout and binding are identical, which suggests that they once formed part of the same manuscript.
3. The disruption in the order of the folios, perhaps as a result of the separation and binding, is visible in both manuscripts, although it is noted by Sezgin only for the second treatise.
4. Some folios have been lost entirely, since the beginning and the ending of *On my own Books* are both missing from the Arabic. We have lost about a page and a half in Greek in Müller's edition from the beginning, but only a few lines at the end.

The first task was the restoration of the correct order of the surviving leaves.[13] The present first folio, fol. 22, must be placed between folios 31 and 32. How did the displacement occur? The Meshed manuscript is composed of ternions and quinions, all signed in the top right hand corner of the verso of the last leaf, and in the bottom left hand corner of the first leaf. They are divided as follows: 1] fols 1-6 = ternion 1: 2] fols 7-16 = quinion 1: 3] fols 17-26 = quinion 2: 4] fols 27-35 = quinion 3, with a missing folio: 5] fols 36-41 = ternion 2. Our text begins on folio 23, but the beginning of the text is now lost, and it is clear that its opening was once written on the verso of folio 22. But what is found there now does not correspond to what we have lost, and the present folio was put there in error to fill in an existing gap. Its original place was between the present folios 31 and 32, with the present folios 31 and 22 forming the central bifolium of the 3rd quinion of our tract. After it had fallen out, folio 22 was wrongly restored after 21 to form the central leaf of the 2nd quinion, where the middle bifolium had also fallen out. This explains why quinion 3 appears today with a missing folio, folios 27-35 instead of 27 to 36.

The loss of the opening leaf deprives us of the precious indication of the title and name of the translator. This presents a difficulty since we know of a second version circulating in the ninth century that was not by Ḥunain, and of which he appears to have been ignorant. The

13 Folios 1r-21v contain another Galenic treatise, *The soul's dependence on the body (Quod animi mores corporis temperamenta sequantur*) in the Arabic translation by Stephanos b. Basil. See H. H. Biesterfeldt, *Galens Traktat 'Daß die Kräfte der Seele den Mischungen des Körpers folgen*, Abhandlungen für die Kunde des Morgenlandes XL.4 (Wiesbaden 1973), who was unaware of the Meshed Ms. The last two folios, fols. 39-40, contain an unidentified Arabic text.

Arab historian al-Ya'qubi cites this treatise several times in his *Chronicle*.[14] But his quotations employ a different technical vocabulary, especially in anatomy, from that in the Meshed Ms., which, as Ivan Garofalo has confirmed for me, is that usually used by Ḥunain.[15] The second Meshed Ms., Rida, tibb 5214/1, presents exactly the same features of vocabulary, and we have argued that the two manuscripts once formed part of the same book, which was divided in two to give a roughly equal number of folios, 41 and 46. In the second Meshed Ms, the translator of *On the Order of my own Books* is clearly said to be Ḥunain, who we know from his *Risāla,* nos. 1-2, to have translated both works. There is thus no real reason to doubt that Ḥunain was the author of the Meshed version of *On my own Books.*

The Meshed manuscript, despite its lacunae, is of immense value for the editor, from many points of view.

1. Titles. The Greek text of *On my own Books* is divided into chapters, each with its own chapter heading. That these headings are old is confirmed by the Arabic, which allows us to put back several headings that are lost in Greek. The titles in Arabic are written within a continuous text, but they are marked out by a much brighter ink (probably red, but this is not clear from the photograph).

The asterisks on the following list of chapter divisions show where the headings are transmitted only in Arabic, which allows us to restore three titles absent from the lacunose Greek text:[16]

Greek (ms. A)	**Arabic (Ḥunain's version)**
Title	Title and opening are missing
1 93,17 M.	1 fol. 23v.
2 97,3 M.	2 fol. 25v.
lacuna 98,11 M.	3 fol. 26r*
3 102,20 M.	4 fol. 28r.
lacuna (anatomy)	5 fol. 31v*
lacuna (anatomy)	6 fol. 22r.*
4 109,4 M.	7 fol. 22v.
5 110,1 M.	8 fol. 32r.
6 111,9 M.	9 fol. 32v.
7 114,8 M.	10 fol. 34r.
8 115,1 M.	11 fol. 34v.
9 115,5 M.	12 fol. 34v.
10 115,16 M.	13 fol. 35r.
11 115,19 M.	14 fol. 35r.
12-16	no divisions and titles
17	lost in Arabic, along with the end of 16

2. New information. Most significant of all, Ḥunain's translation has preserved an important passage relating to Galen's anatomical treatises that we can no longer find in the

14 M. Klamroth, 'Über die Auszüge aus griechischen Schriftstellern bei Al-Yaq'ubi. II. Die übrigen Ärzte', *ZDMG* 40 (1886) 612-38, esp. 625 ff., no. 25.

15 Cf. I. Garofalo, 'Note filologiche sull'anatomia di Galeno', *ANRW* II 37,2 (1994), 1790-833, esp. 1823 ff.

16 The headings are not printed in K.

Greek.[17] In chapter 3 of *On my own Books*, when dealing with anatomy, Galen summarises the contents of the 20 books of anatomy written by Marinus, who in Galen's eyes had revitalised the study of anatomy around 129 AD after centuries of neglect.[18] He tells us that he himself abridged the 20 books of Marinus into four, and he produced a similar abridgement of Lycus' anatomical treatise into two books. Galen's report of his summary of Marinus' treatise on anatomy is our only information about what constitutes the oldest complete treatise on anatomy and physiology known from the ancient world. Unfortunately, Galen's description breaks off in Greek half way through his analysis of book 19, leaving us completely ignorant about the contents of the last book. The loss of the bifolium in A, which is responsible for this, also deprived us totally of Galen's description of Lycus' treatise.[19] This loss is all the greater since we do not know the precise number of books that made it up. This is what makes the new evidence of the Arabic translation particularly precious. It tells us first of all that Marinus devoted his final book 20 to a description of the nerves coming from the cerebellum. But even more importantly, it tells us that Lycus' treatise was almost as large as that of Marinus, in fact in nineteen books. It dealt with both dissection and vivisection, and included a mention of the dissection of a uterus containing a foetus. Galen abridged it in two parts, the first covering books 1 to 9; the second books 10-19.

The contents of the individual books were as follows:

Book 1
1: the brain
2: the nerves issuing from the brain and from the membrane of the brain[20]
3: the nerves issuing from the brain[21]
4: the eye
5: the head of the pharynx[22]
6: the lung after death
7: the lung in life
8: the heart (?) [23]
9: the diaphragm

Book 2
10: the liver
11: the omentum and spleen
12: the kidney
13: the bladder and penis

17 An English translation from the Arabic is included in the Appendix, see below pp. 16-18.

18 M. Grmek, D. Gourevitch, 'Aux sources de la doctrine médicale de Galien: l'enseignement de Marinus, Quintus et Numisianus', *ANRW* II 37,2 (1994) 1491-528.

19 Lycus of Macedon, a pupil of Quintus, moved to Rome, but seems to have died before or shortly after Galen's arrival there in 162 AD. His opinions on anatomy, medicine, and Hippocratic interpretation are roundly attacked by Galen, see F. E. Kind, s.v. Lykos, *RE* 13,2 (Stuttgart 1927) 2408-17.

20 What Lycus was describing is not clear in Ḥunain's translation. The word used in Arabic to describe the membrane is the same used to describe the diaphragm. Did Lycus distinguish between cerebrum and cerebellum, or between pia and dura mater, or between different ventricles of the brain?

21 The scribe may, perhaps, have made a mistake, writing the same sentence twice.

22 Probably, as Vivian Nutton suggested to me, some word (or words) have been omitted after 'head'.

23 The relevant word is missing in the manuscript, which has a blank; I have adopted the suggestion of V. Nutton.

14: the uterus
15: the dissection of a uterus containing a foetus
16: a living child[24]
17: a dead child
18: the testicles
19: the muscles

This new information, by restoring to us the contents of treatises otherwise lost, allows us to write a new chapter in the history of anatomy and medicine.

But the interest of the Arabic text does not stop here, since it allows us to see the lists of his own anatomical treatises that Galen included in the subsequent passage; and these are followed by his writings on physiology and therapeutics. Here the rewards in terms of new discoveries are less, since most of the titles were already known to us. However, Galen mentions at least two tracts whose title appears to be new: the first *On the Science of Anatomy*; the second *On the Faculties of the Soul*,[25] which would seem worth adding to the already long list of books by the doctor from Pergamum.

Galenic Book Numbers

As well as bringing new information, Ḥunain's translation raises a series of problems. In particular, the number of books constituting a particular treatise can undergo a variety of changes. In the first place, Ḥunain's translation presents several differences between it and our other sources, notably his own *Risāla*. To give a few examples, Ibn abi Usaibi'a reports that Lycus wrote 17 books on anatomy, but Ḥunain's translation gives 19, a figure that has every chance of being right. On the other hand, Ḥunain uses a problematic form of the dual to describe the treatise *On the Differences between the homoeomerous Parts of the Body*, whereas in the *Risāla* he describes it as comprising a single book, and the last figure is confirmed by Gotthard Strohmaier, its *CMG* editor.[26]

The problems are even greater where one is dealing with treatises that are lost in Arabic as well as in Greek. So Galen's treatise *On Anatomy in Hippocrates* has six books according to Ḥunain's translation of *On my own Books*, but only 5 according to the *Risāla*, the latter being almost certainly wrong.[27]

A second and more complex example concerns the two references to *On the Opinions of Hippocrates and Plato* in *On my own Books*. The first comes in a passage lost in Greek but preserved in Arabic, the other, at the end of chapter 13, is available in both Greek and Arabic. In both instances the Arabic translation specifies that the tract contained ten books,

24 Hunain uses different words for 'foetus' and for 'child'.

25 Unless Galen is talking in general terms without having a specific treatise in mind.

26 G. Strohmaier, ed., *Galeni De partium homoeomerium differentia, CMG* Supplementum Orientale III (Berlin 1970). The editor (32, note 8) offered two explanations for the absence of this treatise from *On my own Books* and from Ḥunain's list of works omitted from it: an oversight by Ḥunain (his preferred explanation) or its presence in Ḥunain's text of *On my own Books*.

27 Not least, because in a passage in *On my own Books* 1: 95,13 M. = XIX 14,4 K., Galen says that this treatise had six books.

whereas the Greek of chapter 13 says that it had only 9.[28] Philip De Lacy, in his *CMG* edition, relying on Ḥunain's testimony in the *Risāla* and on that of Rhazes in the *Kitab al-Hawi*, raised the possibility of a tenth book, lost in Greek, and suggested that it was written after *On my own Books*, since it was not mentioned there.[29] Now, as we have seen, Ḥunain in his translation already specifies ten books, not nine, so De Lacy's suggestion cannot be right. But are we then dealing with a mistake in the Greek tradition, or with a later alteration by Ḥunain? And, if we think that Ḥunain was right, when was the tenth book written, since it must precede *On my own Books*? For the moment, the question remains open.

As we have seen, the Arabic tradition represented by Ḥunain's version raises as many questions as it solves. Some chapters, especially chapters 9 and 12, devoted to Galen's writings on the Empiricists and on ethics, contain substantial differences between Ḥunain's Arabic and the Greek tradition as it has come down to us. And if that Greek often is clearly corrupt, the Arabic is, for its part, often not much better. In such cases, the more difficult the alternatives, the greater the need for prudence.

APPENDIX

A translation of the newly discovered section of 'On my own Books' from the Meshed MS

fol. 31r.

In Book 19, he deals with the nerves that come from the brain, with smell and the origin of that sense, with the optic nerves (called by Herophilus and Erasistratus 'canals'), and lastly[30] the nerves coming from the ears. In Book 20 he describes the nerve coming from the lower part of the brain and the spinal cord that comes from it. As for the two books in which I abridged the treatise of Lycus on anatomy,[31] I included in the first book abridgements of nine books of Lycus on anatomy. In the first he discusses the brain, in the second the nerves coming from the brain and from the membrane of the brain, and in the third the nerves coming from the brain. In Book 4 he mentions the eye, in Book 5 the head of the pharynx,[32] in Book 6 the lung in death, in Book 7 the lung in life, in Book 8 < >,[33] and in Book 9 the diaphragm. In the second of my two books in which I abridged Lycus' work, I abridged ten books. The first considers the liver; the second the membrane which covers the stomach

28 *On my own Books* 13: 122,17-18 M. = XIX 46, 19 K. The *Risāla*, no. 46, also gives this treatise ten books.

29 P. De Lacy, ed., *Galeni De placitis Hippocratis et Platonis, CMG* V 4,1,2, 44-45: 'This hints at the possibility that Galen added this tenth books later in life and after *De libris propriis*, where he explicitly indicates the number of books in PHP to be nine. One must further assume that this tenth book, like the beginning of Book I, has been lost in the Greek textual tradition.'

30 Here begins the lacuna in Greek, 108,14 M. = XIX 34 K.

31 Ibn Abi Usaibi'a described Lycus' work as being in 17 books (see Bergsträsser's *apparatus criticus* to the *Risāla*, no. 23), but this error can now be corrected from this translation.

32 For the textual problem, see above, note 22.

33 See above, note 23.

and is called *epiplon*, as well as the spleen; the third the kidney; the fourth the bladder and penis; the fifth the uterus. In the sixth is described the dissection of a uterus of a dead woman in which there is a foetus.[34] In the seventh he discusses the living child, and in the eighth the dead child; the ninth deals with the testicles, and the tenth with the muscles (fol. 31v.). I also composed some treatises whereby those who made use of them could develop their anatomical knowledge, including two books of *Controversies in Anatomy*,[35] a single book *On the Anatomy of dead Subjects* and two on *Vivisection*.[36] I have also written likewise two books on *The Differences between the homoeomerous Parts of the Body*.[37] I have also said that I wrote six books on *Anatomy in Hippocrates*, and three on *Anatomy in Erasistratus*.[38] I also composed a commentary in four books on *What Lycus did not know about Anatomy*.[39] After this I wrote two books on *Differences with Lycus on Anatomy*, and a single book *On the Science of Anatomy*, so that you might learn from them about the composition of all the parts of the body.[40] This book was followed by others that I wrote on the faculty of each of these parts and on their usefulness.[41] That is what I entitled them.

Chapter 5 Mention of the books in which I explained what could be seen in a dissection as far as concerned the faculty of the parts and their usefulness.

I have already mentioned that I composed three books on *The Movement of Thorax and Lung* and two books on *The Causes of Breathing*.[42] These are followed by four books on *The Voice*, and I also composed on this topic two books on *The Movement of Muscles* and what I wrote on *The Faculties of the Soul*.[43] I also composed three books on the natural faculties, which I entitled *A Work on the natural Faculties*.[44] I further wrote on these topics a single book in which I criticised a novel (fol. 22 r.) opinion on the explication of blood:[45] a single

34 It is unclear whether this was a Caesarean section or a discovery made during a *post mortem* dissection.

35 Περὶ τῆς ἀνατομικῆς διαφωνίας = *Risāla*, no. 24, now lost in Greek. Ibn abi Usaibi'a added 'among the ancients'.

36 Περὶ τεθνεώτων ἀνατομῆς, Περὶ ζώντων ἀνατομῆς. Both these treatises, *Risāla*, no. 25-26, are lost in Greek, but see I. Ormos, 'Bemerkungen zur editorischen Bearbeitung der Galenschrift 'Über die Sektion toter Lebewesen', in J. Kollesch, D. Nickel, eds, *Galen und das hellenistische Erbe*. Sudhoffs Archiv Beiheft 32 (Stuttgart 1993) 165-72.

37 Περὶ τῆς ὁμοιομερῶν σωμάτων διαφωνίας = *Risāla*, no. 33. See above, note 26.

38 Περὶ τῆς Ἱπποκράτους ἀνατομῆς, Περὶ τῆς Ἐρασιστράτου ἀνατομῆς = *Risāla*, nos. 27-28 (cf. above, note 27). Both are lost.

39 Περὶ τῶν ἀγνοηθέντων τῷ Λύκῳ κατὰ τὰς Ἱπποκράτους ἀνατομάς = *Risāla*, no. 29: now lost.

40 Hunain did not mention these books, both of which are lost, in his *Risāla*.

41 *On the usefulness of parts*, III 1-933 K., unless Galen is talking in general terms. An identification also with *On the natural faculties*, II 1-214 K., would seem to be excluded by Galen's description of this tract, below, note 44.

42 Περὶ θώρακος καὶ πνεύμονος κινήσεως, Περὶ τῶν τῆς ἀναπνοῆς αἰτιῶν = *Risāla*, nos 36-37. Both are lost, although a summary of the latter may be found at IV 465-69 K.

43 Περὶ φωνῆς = *Risāla*, no. 38, now lost, see H. Baumgarten, *Galen, Über die Stimme. Testimonien der verlorenen Schrift* Περὶ φωνῆς, Diss. (Göttingen 1962). Περὶ μυῶν κινήσεως = *Risāla*, no. 39: IV 367-464 K. The identity of the last work is unclear.

44 Περὶ φυσικῶν δυνάμεων = *Risāla* 13: II 1-214 K.

45 Unknown, but possibly identical with the tract mentioned by Ḥunain, *Risāla*, no. 40: *On a new theory of distinguishing urines*: Πρὸς τὴν καινὴν δόξαν περὶ τῆς τῶν οὔρων διαφορᾶς. But the text in one or other of these places is clearly corrupt.

book on *The Usefulness of the Pulse*,[46] and a single book on *The Usefulness of Breathing*;[47] a single book in which I showed that blood is contained naturally in the arteries;[48] and a single book on *The Properties of Purgatives*.[49] I also wrote a single book in which I described the words that one most know in investigating the faculties of the rational soul, which are imagination, intellect, and memory;[50] <and a treatise on> the faculties which are at the very foundation of the body, in ten books, which I called *On the Opinions of Hippocrates and Plato*.[51] Finally come the seventeen books that I composed on *The Usefulness of Parts of the Body*.[52]

Chapter 6. Description of what one must observe before considering therapy

The first thing that must be considered in this aspect of science is the book I wrote on *Elements according to Hippocrates*.[53] In it I showed that the hot, cold, wet and dry are the universal elements that form bodies susceptible of generation and corruption. We called them elements in accordance with their substances, viz. earth, fire, air, and water. In fact the elements in the body of humans and in the bodies of animals with blood are blood, phlegm and the two sorts of bile, yellow and black. After the single book *On Elements* comes the work *On Mixtures*, which is in three books.[54] Book 3 bears on the mixtures of drugs, and for this reason we should follow this with the tract on *The Properties of Simples*, in eleven books.[55] But one may postpone reading Books 1 and 2 of *On Mixtures* until after the tract on *The Properties of Simples*. There follow, after Books 1 and 2 of *On Mixtures*, a book I wrote on *The best Constitution*; a single book on *The best State of the Body*; and the single book on *The Irregularities of an unbalanced Temperament*:[56] and, after that, a single book on *The Differences between Diseases*....[57]

Paris

46 Περὶ χρείας σφυγμῶν = *Risāla* no. 41: V 149-80 K.

47 Περὶ χρείας ἀναπνοῆς, II 470-511 K. Not mentioned in the *Risāla*.

48 Εἰ κατὰ φύσιν ἐν ἀρτηρίαις αἷμα περιέχεται = *Risāla*, no. 43: IV 703-36 K.

49 Περὶ τῆς τῶν καθαιρόντων φαρμάκων δυνάμεως = *Risāla*, no. 44. A fragment is printed at XI 323-42 K.

50 Not so far identified.

51 Περὶ τῶν Ἱπποκράτους καὶ Πλάτωνος δογμάτων = *Risāla*, no. 46: V 181-805 K. See also, above, n. 28.

52 Περὶ χρείας μορίων = *Risāla*, no. 49: III 1-933 K.

53 Περὶ τῶν καθ' Ἱπποκράτην στοιχείων = *Risāla*, no. 11: I 413-508 K.

54 Περὶ κράσεων = *Risāla*, no. 12: I 509-694 K.

55 Περὶ δυνάμεως τῶν ἁπλῶν φαρμάκων = *Risāla*, no. 53: XI 379-XII 377 K.

56 Περὶ τῆς ἀρίστης κατασκευῆς τοῦ σώματος = *Risāla*, no. 50: IV 737-49 K.; Περὶ τῆς εὐεξίας = *Risāla*, no. 51: IV 750-56 K.; Περὶ ἀνωμάλου δυσκρασίας = *Risāla*, no. 52: VII 733-52 K.

57 End of the lacuna in Greek: 108,15 M.

DIVISION, DISSECTION, AND SPECIALIZATION: GALEN'S 'ON THE PARTS OF THE MEDICAL TECHNE'

HEINRICH VON STADEN

1. Introduction

Karl Gottlob Kühn shaped the modern reception of Galen not only by producing the sole 'complete edition' of Galen's works published since the seventeenth century but also by omitting previously known and printed Galenic texts from his *Claudii Galeni Opera Omnia* (1821-33). Among such omitted texts is *On the Parts of the Medical Techne*. The suppression of this text, as of several others surviving only in Latin, was not due to inadvertence or ignorance on Kühn's part. In the remarkable *Historia literaria Claudii Galeni* (by Johann Christian Gottlieb Ackermann) that Kühn published in 1821 as the 249 page introduction to volume I of his edition,[1] *On the Parts of the Medical Techne* is listed as Galenic treatise no. 103 and is identified as extant only in Latin. Furthermore, it is described as previously published in René Chartier's seventeenth-century edition – on which Kühn's edition is largely based – and in a sixteenth-century Juntine edition which assigned it to the spurious Galenic works.[2] Accepting the Juntine verdict, Ackermann enumerated the treatise among forty-four 'clearly spurious books' transmitted with Galen's authentic works.[3] Kühn was in complete agreement with Ackermann's view, as he soon made clear.

Two years after the publication of the first volume of his edition Kühn took up the question of the authenticity of *On the Parts of the Medical Techne* at greater length. In an *Universitätsprogramm* of the University of Leipzig he argued that the treatise was spurious,[4] expressing the suspicion that the Latin 'translation' – to his knowledge the only extant version – of the treatise was a forgery by Niccolò da Reggio, the famous early fourteenth-

1 Ackermann (1756-1801) compiled his *Historia literaria* on the basis of the fourth edition of Johann Albert Fabricius' famous *Bibliotheca Graeca* (1790), revised and augmented by Gottlieb Christoph Harles (1738-1815). See Kühn, *Galeni Opera Omnia*, I xvii: 'Historia Literaria Claudii Galeni, conscripta ab Jo. Chr. Gli. Ackermanno, ex Jo. Alb. Fabricii Bibl. Graec. ex ed. Gli. Cph. Harles, vol. V. 377-500. desumpta et hinc inde aucta et emendata'.

2 Kühn, *Galeni opera* (n. 1 above) I clx: '103. *De partibus artis medicae*. Exstat in ed. Chart. to. II. 282. in VII. Junt. in class. spurior. f. 16'.

3 Kühn, *Galeni opera* (n. 1 above) I clviii: 'Libri manifeste spurii, qui inter Galeni opera recepti leguntur ...' The editor of the fourth Juntine edition (Venice 1565), A. Gadaldinus, steered clear of judgments such as *manifeste spurium* and instead offered a more balanced judgement about the authorship of *On the Parts of the Medical Techne* (even though he still included it in the volume containing τὰ νόθα): 'Liber, qui nisi Galeni fuerit, eo tamen auctore dignus uidetur'.

4 C. G. Kühn, 'De loco Celsi in praefat. 3 ed. Targ. noviss. male intellecto exponitur,' *Universitätsprogramm Leipzig* (1823); repr. in C. G. Kühn, *Opuscula academica et philologica*, vol. 2 (Leipzig, 1828) 225-31.

century translator of Galenic texts from Greek into Latin.[5] Although Kühn did not hesitate to include – in a later volume of his edition of Galen's *Opera omnia* – many of the forty-four works labelled 'clearly spurious' in Ackermann's prolegomena,[6] *On the Parts of the Medical Techne* remained banished from Kühn's edition. As a consequence this remarkable treatise, so well known in the Renaissance and repeatedly published between 1490 and the late seventeenth century, almost disappeared from view for close to a century after its exile from Kühn's *Cl. Galeni Opera Omnia.*[7] Only in 1911 did Hermann Schöne, an acute reader of Galen, draw more than cursory attention to the text again.[8] Schöne defended its authenticity and published a critical edition of its Latin version. The subsequent discovery of a ninth-century Arabic translation of the text at one stroke demolished Kühn's conclusion that the work was a forgery by Niccolò da Reggio, but it still left many questions, including that of the authenticity of the work, unanswered. Even after the publication of the Arabic translation, the treatise remained relatively neglected in modern Galenic scholarship, and no comprehensive analysis of its argument has been attempted. The present contribution tries to re-animate interest in this remarkable text, which presents far more puzzles and problems than could be discussed adequately in this brief article. Following introductory remarks on the transmission of the text (Part 2) and on the question of its authenticity (Part 3), I shall focus on Galen's sustained use of dissection as a methodological model for the division of

5 Kühn, *Opuscula* (n. 4 above) II 230, relied in part on erroneous chronological assumptions to argue for the inauthenticity of *On the Parts of the Medical Techne*: he suggested that Niccolò composed the forgery at the prompting of Venetian publishers who wished to profit from Galen's great authority, when in fact the earliest printed edition of this Galenic text dates to 1490, well over a century after Niccolò's death. On Niccolò of Reggio (c. 1280-1350), who worked at the Angevin court as a physician and translator of Greek medical texts into Latin, see R. Sabbadini, 'Le opere di Galeno tradotte da Nicola de Deoprepio di Reggio,' in *Studi storici e giuridici dedicati ed offerti a Federico Ciccaglione*, 2 vols. (Catania, 1909-10) II, pt. iii, 15-24; L. Thorndike, 'Translations of Works from the Greek by Niccolò da Reggio (c. 1308-1345)', *Byzantina Metabyzantina* 1.1 (1946) 213-35; R. Weiss, 'The translators from the Greek of the Angevin Court of Naples,' *Rinascimento* 1 (1950) 195-226; R. J. Durling, 'A chronological census of Renaissance editions and translations of Galen', *JWI* 24 (1961) 230-305 (233); I. Wille, 'Überlieferung und Übersetzung. Zur Übersetzungstechnik des Nikolaus von Rhegium in Galens Schrift De temporibus morborum', *Helikon* 3 (1963) 259-277; G. Baader, 'Mittelalterliche Medizin im italienischen Frühhumanismus,' in G. Keil (ed.), *Fachprosa-Studien: Beiträge zur mittelalterlichen Wissenschafts- und Geistesgeschichte* (Berlin 1982) 204-54; and M. H. McVaugh, below, pp. 153-63.

6 In vol. 19 of his edition Kühn included not only some genuine works but also the following works listed among the *Libri manifeste spurii* by Ackermann (and still widely regarded as pseudo-Galenic): *De historia philosophica, Definitiones medicae, Quod qualitates incorporeae sint, De humoribus, Praesagitio omnino uera expertaque, De uenae sectione, Prognostica de decubitu ex mathematica scientia, De urinis, De pulsibus ad Antonium disciplinae studiosum ac philosophum, De affectuum renibus insidentium dignotione et curatione, De ponderibus et mensuris doctrina*, and *De succedaneis*.

7 After Kühn it was briefly mentioned as a spurious work by L. Choulant, *Handbuch der Bücherkunde für die ältere Medizin*, 2nd ed. (Leipzig 1841) 108. C. Daremberg, *La médecine. Histoire et doctrines* (Paris, 1865) 485 more cautiously labelled it 'attributed to Galen'. J. Ilberg, 'Über die Schriftstellerei des Klaudios Galenos IV', *RhM* 52 (1897) 591-623 (603 n.), however, recognized it as authentic.

8 H. Schöne, *Galenus de partibus artis medicativae, eine verschollene griechische Schrift in Übersetzung des 14. Jahrhunderts. Festschrift der Universität Greifswald zum Rektoratswechsel am 15. Mai 1911* (Greifswald, 1911). This edition of the Latin text was reprinted with minor editorial modifications by J. Kollesch, D. Nickel, and G. Strohmaier in *CMG, Suppl. Orientale* 2 (Berlin, 1969) 115-29. Unless otherwise noted, all references below to *De partibus artis medicativae* [hereafter = *PAM*] will be to this *CMG* edition, which also includes the Arabic text with an English translation (see nn. 10, 19 below).

the medical *techne* into its various branches (Part 4), and on his views on the relation between division, specialization, and urbanization (Part 5).

2. Transmission

As far as we know, the text of *On the Parts of the Medical Techne* (hereafter = *PAM*) no longer is extant in Greek, but it seems to have been known in at least two closely related Greek versions during the Middle Ages. One branch of the Greek manuscript tradition is indirectly represented by a ninth-century Arabic translation that survives in an eleventh-century manuscript of the Aya Sofya collection in Istanbul (cod. Aya Sofya 3725, dated to the year 1063). It is on this manuscript, to which Hellmut Ritter and Richard Walzer first drew attention,[9] that Malcolm Lyons' critical edition and translation of the Arabic version are based.[10] The identity of the Arabic translator remains uncertain. Ḥunain ibn Isḥāq reports in his *Risāla* that he translated this Galenic work into Syriac for a man called 'Alī, known as al-Fayyūm'.[11] A gloss entered in the *Risāla* adds that 'Ḥunain translated more than half of this [Galenic] work about two months before his death and [his son] Isḥāq finished it; the translation was made into Arabic'.[12] Lyons observed, however, that the Arabic translation appears stylistically inferior to the best Arabic translators, among whom 'the place of Ḥunain and Isḥāq is unquestioned', and he offers several examples of 'oddity of expression, clumsiness of construction, and ambiguity.'[13]

In the early decades of the fourteenth the century the Greek physician Niccolò da Reggio translated the treatise into Latin from a Greek manuscript that apparently represents a different textual tradition (see n. 19 below). Since no preface to his translation of *PAM* has survived, we regrettably do not have the kind of information that Niccolò provides in the extant prefaces to some of his other Latin translations of Galen (for example, an account of the circumstances that prompted him to undertake the translation or a direct first-person confirmation of his role as translator).[14] But the extant colophon to Niccolò's rendering of

9 H. Ritter, R. Walzer, 'Arabische Übersetzungen griechischer Ärzte in Stambuler Bibliotheken,' *Sitzungsberichte der Preußischen Akademie der Wissenschaften*, phil.-hist. Kl. (Berlin 1934) 801-46 (816 no. 27). See also M. Ullmann, *Die Medizin im Islam*. Handbuch der Orientalistik, Ergänzungsband VI. 1 (Leiden/Cologne 1970) 52 no. 69.

10 *CMG Suppl. Orient.* 2 (Berlin 1969) 22-29 [hereafter = *PAM* (see n. 8 above); Lyons' English translation from the Arabic is cited as: Lyons, *PAM*]. See n. 19 below.

11 G. Bergsträßer, 'Ḥunain ibn Isḥāq über die syrischen und arabischen Galen-Übersetzungen', *Abhandlungen für die Kunde des Morgenlandes* 17.2 (Leipzig 1925) no. 61; translation by Lyons, *PAM* (n. 10 above) 8.

12 Bergsträßer, 'Ḥunain' (n. 11 above) no. 61; translation by Lyons, *PAM* (n. 10 above) 8. In recension B, however, this gloss states unequivocally that Ḥunain translated the text into Arabic; see G. Bergsträsser, 'Neue Materialien zu Ḥunain ibn Isḥāq's Galen-Bibliographie', *Abhandlungen für die Kunde des Morgenlandes* 19.2 (Leipzig 1932) 34-35: 'Dann habe ich begonnen, einen Teil davon *ins Arabische* zu übertragen. Isḥāq hat seine (des Buches) Übersetzung vollendet nach dem Tode seines Vaters' (italics added).

13 Lyons, *PAM* (n. 10 above) 9. Lyons nevertheless emphasizes that such features do not constitute conclusive evidence about the authorship of the extant Arabic translation.

14 In Niccolò's prefaces he remarked, for example: 'Ego Nicolaus de Regio medicus fidelis et subditus uester iuxta mandatum uestrum de greco idiomate in latinum transtuli nihil addens, minuens uel permutans' (addressing King Robert I of Anjou in the preface to his translation of Galen's *De compositione medicamentorum secundum locos*). Also: 'libros Gal. duos ... de greco ydiomate in latinum transtuli Vestre magnitudini offerendos' (in the preface to his translations of the spurious *De uenae sectione* and of Galen's *De utilitate respirationis*, again addressed to Robert, but before Robert became King in 1309). And: '...me, Nicolaum de Regio de Calabria, medicum, cum instantia requis-

PAM corresponds to the colophons of his other translations: *Explicit libellus Galieni de partibus medicatiue translatus a magistro Nicolao de Regio de Calabria.*[15] Doubts have been expressed in recent decades about the authorship of some of the translations attributed to Niccolò, but in this case there is no compelling reason to question the veracity of the colophon. The textual evidence is further complicated by the fact that no manuscript containing Niccolò's translation has been discovered and that our knowledge of the Latin text therefore depends exclusively on Renaissance editions of Galen's works. In the absence of any Greek or Latin manuscript, the discovery of the extant Arabic version is all the more significant, even if Niccolò's more literal translation at times makes the reconstruction of the original Greek easier than does its Arabic counterpart.

Kühn apparently failed to notice that two versions of the Latin translation of *PAM* are extant. The first, which Hermann Schöne quite plausibly took to be Niccolò's own, unadulterated rendering, appears in the Latin editio princeps of Galen's works, edited by the physician Diomedes Bonardus and published by Philippus Pincius in Venice in 1490. The same Latin version was included in other early Latin editions of Galen's collected works, for example in the edition by Rusticus Placentinus (published by J. de Burgofranco in Pavia, 1515-16), in the Juntine editions of 1528 and 1541-42 (the latter edited by J. B. Montanus), in the editions published in Basel by H. Froben and N. Episcopius in 1541 and 1549 (edited, respectively, by H. Gemusaeus and J. Cornarius), in C. Gesner's edition published by J. Frellon in Lyons in 1549-51, and in the Juntine edition of 1550 edited by A. Gadaldinus.[16] This 'original' version of the Latin translation therefore enjoyed wide dissemination in the first half of the sixteenth century.

By 1541, however, a second version of the Latin text began to appear. It is a smoother, more polished version, adapting Niccolò's literal translation to the ideals of Renaissance Latin. This in part radically revised version, by the Venetian humanist Vettore Trincavelli, first appeared anonymously in the Latin edition of Galen's collected works by A. Ricci and Trincavelli, published in Venice 1541-45 (ex off. Farrea), and then in 1552, with explicit attribution to Trincavelli, in a Latin edition of four Galenic works published by G. Rouille.[17] In Trincavelli's adaptation the above-mentioned colophon identifying Niccolò as the translator was omitted, as were Niccolò's chapter headings (which had taken the form of introductory summaries). In subsequent sixteenth-century Latin editions of Galen's collected works, and at least until René Chartier's edition of 1638 (*Operum Hippocratis Coi et Galeni Pergameni,* tomus II), Trincavelli's stylistically more agreeable revision supplanted the early printed version of Niccolò's translation. Trincavelli's text was republished, for example, by A. Gadaldinus in his later Juntine editions (1556 and 1565), and likewise by C. Gesner in his edition of 1561-1562 (published in Basel by H. Froben and N. Episcopius). As indicated above, Gesner and Gadaldinus had previously adopted Niccolò's version of the Latin text (in

iuistis et monuistis, ut ipsum a principio usque ad finem transferrem uobis de greco ydiomate in latinum' (from Niccolò's preface to his translation, completed in 1317, of Galen's *De usu partium*); Schöne, *Galenus* (n. 8 above) 7-10 (nos. 18, 8, and 22). See also Schöne 6-11 for a useful collection of other prefaces and colophons that frame Niccolò's extant translations of Galenic texts.

15 Schöne, *Galenus* (n. 8 above) 38, ll. 377-78.

16 See Durling, 'Census' (n. 5 above) 288 (no. 81a).

17 See Durling, 'Census' (n. 5 above) 280 (no. A7b and 270 [1552.2]).

their editions of 1549-51 and 1550, respectively), but they now both joined the Trincavelli bandwagon. J. B. Rasarius also opted for the 'modern' Latin version in his edition of 1562-63 (Venice: V. Valgrisi), as did H. Mercurialis, H. Costaeus, and Fabius Paulinus in their Juntine editions of, respectively, 1576-77, 1586, and 1596-97.[18] So complete did the eclipse of Niccolò's version by Trincavelli's revision become that Niccolò's translation does not seem to have been reprinted for more than 350 years after the mid-sixteenth century (i.e. until Schöne's edition of 1911). The consequences were significant: as Schöne pointed out, Kühn committed the fatal error of basing his deprecatory remarks about Niccolò's supposedly incompetent 'forgery' and 'un-Galenic style' exclusively on Trincavelli's adaptation of Niccolò's translation, even though Niccolò's original version was readily available to Kühn in the University Library of Leipzig, namely, in a copy of Bonardus's Latin edition of Galen's works (Venice 1490).

The many discrepancies between Niccolò's literal Latin translation and the Arabic translation (noted in the *CMG* edition) are not all attributable to the different demands made on the translator by Arabic and by Latin, and it therefore seems almost certain that Ḥunain's Greek manuscript did not belong to the same branch of the transmission as the manuscript translated by Niccolò.[19] On these two lines of indirect transmission – represented, respectively, by a single eleventh-century Arabic manuscript and by Renaissance printed editions of a Latin translation – our knowledge of *PAM* depends.

3. Authenticity

The treatise is not mentioned in Galen's *On my Own Books* or in his *On the Order of my Own Books*, and no unequivocal allusion to it appears in Galen's other works, not even in the 'catalogue' of his works presented at the end of his *Ars Medica* (a work which has much in common with *PAM*). Small wonder, then, that Kühn, who was unaware of the existence of an Arabic translation – but who at times was also notoriously careless and mistaken about details – shared the view of a number of sixteenth-century editors that the treatise is inauthentic (even though René Chartier had presented it as an authentic introductory work, including it in his second volume among Galen's 'τὰ εἰσαγωγικά, *quae in artem medicam introducunt*').

In his arguments against the authenticity of *PAM* Kühn objected, first, that no Greek manuscript of the work survives and, secondly, that Galen did not mention the work in his two autobibliographic works.[20] But neither of these considerations would in and of itself render the treatise inauthentic. For example, no Greek manuscript of Galen's *Protrepticus* survives, yet its Greek text, as printed ever since the Aldine edition of 1525, has been accepted as authentic,[21] not to mention several of Galen's other genuine works transmitted in mediaeval Arabic and Latin translations of which no Greek manuscripts have been found.

18 See Durling, 'Census' (n. 5 above) 228 (no. 81b).

19 In the *CMG* edition (nn. 8, 10 above) most of the discrepancies between the Latin and Arabic versions of *PAM* are signalled by italics both in the Latin text (pp. 119-129) and in Lyons' English translation from the Arabic (pp. 25-49). A systematic analysis of the nature of these discrepancies, and of their implications for the lost Greek manuscript tradition, remains a desideratum; a Galenic expert whose competence in Arabic is less limited than mine would be better suited to undertake such an analysis.

20 See Kühn, *Opuscula* (n. 4 above) II 225-31.

21 On the transmission of Galen's *Protrepticus* now see V. Boudon, *Galien*, tome II (Paris 2000) 43-79.

As for Kühn's second objection, it is worth recalling that several of Galen's authentic works are not mentioned in his autobibliographical essays, including *On My Own Opinions, On the Instrument of Smell, On the Formation of Embryos, On Antidotes,* and *On Prognosis*. Some of these treatises were written only after Galen had composed his two autobibliographical works, while others perhaps went unmentioned because, as Vivian Nutton has suggested, Galen believed them to be lost due to fires or other circumstances and therefore no longer in circulation; the ostensible purpose of *On My Own Books* was, after all, to establish which of the works circulating under his name were authentic.[22]

A third argument advanced by Kühn against the authenticity of the treatise is that its style is so remote from Galen's abundantly attested style that, even if there were independent evidence of Galenic authorship, one would be no less compelled to conclude that *spurium esse hunc ipsum librum*.[23] Given Kühn's ample familiarity with Galen's genuine works, this judgment is not to be rejected lightly. As indicated above, however, Schöne demonstrated convincingly that Kühn's judgement concerning the style of the text was not based directly on Niccolò's quite literal Latin translation but on Trincavelli's revision thereof, as reproduced in Chartier's edition.[24] Furthermore, Schöne pointed out that Niccolò's literal translation, at least as repeatedly printed from 1490 to 1550, in fact displays stylistic and doctrinal features amply familiar from Galen's extant genuine works. Many such features could be added to those identified by Schöne, but for present purposes a few brief examples will have to suffice.

First, the text displays Galen's well-known habit of frequently referring back to his earlier works that had touched on related issues. In these invaluable cross-references, as throughout the text, there are some divergences between the Arabic and Latin translations, but it seems reasonably certain that *PAM* refers at least twelve times, and perhaps as many as fifteen, to earlier treatises by Galen. The works to which *PAM* makes unequivocal reference include, in the sequence in which they first occur, *On Anatomical Procedures,*[25] *On the Method of Healing,*[26] *On Differences between Diseases,*[27] *Thrasybulus,*[28] *On the Causes of Diseases,*[29] and *Hygiene.*[30] At times the author uses less specific, vaguer expressions in the cross-references, such as 'my books on drugs' and 'with this we have already dealt elsewhere'. The contexts in which these less precise references occur render it likely that they allude to some or all of the following works: *On the Usefulness of Parts, On the Elements According to Hippocrates, On Critical Turning Points, On the Natural Faculties, On the Dissection of*

22 *De libris propriis* (XIX 9.13-10.14 K = *Scr. min.* II 92.2-24 Müller). See V. Nutton, *CMG* V 8,1 (Berlin, 1979) 48-51 on the relevance of this remark for *On Prognosis*.

23 Kühn, *Opuscula II* (n. 4 above) 230.

24 Schöne, *Galenus* (n. 8 above) 14-15.

25 *PAM* 3.3 (pp. 30-31, 121.13, 121.18).

26 *PAM* 4.6 (pp. 34-35, 123.15-16); 7.5 (pp. 44-45, 127.10-11); 9.3 (pp. 48-49, 128.34-129.3).

27 *PAM* 5.2 (pp. 36-37, 123.22-24); 8.5 (pp. 46-47, 128.7-9); 9.2 (pp. 48-49, 128.32-34).

28 *PAM* 7.2 (pp. 42-43, 126.6-7).

29 *PAM* 9.2 (pp. 48-49, 128.32-34).

30 *PAM* 9.3 (pp. 48-49, 128.34-129.3); ch. 5.3 (pp. 36-37, 123.25-26) perhaps also refers to *Hygiene*.

Living Beings,[31] *On the Mixtures and Faculties of Simple Drugs, On the Composition of Drugs According to Places, On the Composition of Drugs According to Kinds,*[32] *On the Differences between Symptoms*, and *On the Causes of Symptoms.*[33] In the Arabic translation a problematic passage refers to a work by Cassius,[34] whereas Niccolò's Latin version of the same passage seems to refer to a work by Galen himself.[35] If the Latin version is correct, Galen here might have alluded to one of the works on Empiricism mentioned in *On My Own Books*, for example his lost treatise in three books *On Disagreements between the Empiricists* (or perhaps his lost work *On Serapion's Against the Sects*).[36]

This web of cross-references to other parts of the Galenic corpus not only is in keeping with Galen's customary practice but also suggests that *On the Parts of the Medical Techne* might be one of Galen's later works (which could in part account for its omission from his two autobiographical essays). Such cross-references obviously do not, in and of themselves, prove the authenticity of any Galenic treatise, but, along with numerous stylistic, thematic, doxographic, and doctrinal affinities with Galen's genuine works (see below), they seem to point in the direction of its authenticity.

Many features of Galen's technical terminology are visible in both the Arabic and Latin translations of *PAM*. The author's repeated references, for example, to the 'perceptible elements' (i.e., as opposed to the four traditional 'natural elements' or 'real elements') is consonant with Galen's distinction between τὰ αἰσθητὰ στοιχεῖα (or τὰ φαινόμενα στοιχεῖα) and τὰ φύσει στοιχεῖα (or τὰ ὄντως στοιχεῖα). Similarly, the author of *PAM* introduces as distinct, separate branches of medicine those that deal with preparatory care, after-care, a 'good condition', old age, cataract couching, and the 'beautification' of patients, faithfully reflecting Galen's uses elsewhere of τὸ παρασκευαστικὸν (*scil.* μέρος τῆς τέχνης) and of the parts called ἀποθεραπευτικόν, εὐεκτικόν, γηροκομικόν, παρακεντητικόν, and κομμωτικόν (see Part 4 below).

Not only nomenclature but also numerous doctrinal details in *PAM* reflect Galen's independently attested views. The systematic distinction in *PAM* between 'homoeomerous parts' and 'instrumental parts' of the body, for example, is consistent with Galen's deployment of the contrast (ultimately derived from Aristotle) between τὰ ὁμοιομερῆ μόρια and τὰ ὀργανικὰ μόρια (or μέρη) τοῦ σώματος. Similarly, the author's tripartite division of

31 In *PAM* 7.4 (pp. 42-43, 126.24-127.3) the author may have in mind some or all of the first five treatises enumerated here, along with *On Anatomical Procedures*.

32 *PAM* 8.6 (pp. 46-47, 128.18): 'As I have shown in my books on drugs' (Lyons); *sicut in libris qui de farmaciis ostensum est* (Niccolò).

33 *PAM* 9.2 (pp. 48-49 in the Arabic translation) = 9.1 (p. 128.26-27 in the Latin version) Galen distinguishes between a work on symptoms of diseases and a work on causes.

34 Lyons, *PAM* 6.6 (pp. 41.37-43.2): 'There is no need for me to give any explanation of this point here as it was clearly shown by Cassius the theoretician in a whole treatise devoted to this subject and the point was not opposed by Serapion the Empiricist. Those Empiricists who wish to learn this should study that treatise'. The Latin version, by contrast, reads as follows (p. 125.22-26): 'quare autem, non instat sermo qui nunc est; sufficienter enim ostensum est per totum unum librum per se. et neque Serapio aliter cognoscebat'. On Cassius see H. von Staden, 'Was Cassius an Empiricist? Reflections on method', in *Synodia. Studia humanitatis Antonio Garzya septuagenario ab amicis atque discipulis dicata*, a cura di Ugo Criscuolo e Riccardo Maisano (Napoli: M. D'Auria Editore, 1997), 939-60.

35 *PAM* 6.6 (p. 125.23-24).

36 *De libris propriis 9* (XIX 38 K = *Scripta minora* II 115 Müller).

the bodily dispositions (διαθέσεις) into diseased, healthy, and neither-of-the-two[37] corresponds to the division in Galen's *Medical Techne* (see Part 4 below).

A further telling affinity between *On the Parts of the Medical Techne* and some of Galen's works of undisputed authenticity is its tendency to enumerate several terminological alternatives and then to claim that 'it makes no difference in the present case' which of them one adopts, for example (in Niccolò's translation): *noticias autem uel notiones uel scientias uel theoremata dicere nihil differt ad presens, sicut neque partes et particulas et portiones*.[38] The corresponding Greek phrases – οὐδὲν γὰρ διαφέρει πρός γε τὰ παρόντα, οὐδὲν πρὸς τὸ παρὸν διαφέρει, οὐδὲν ἔν γε τῷ νῦν διαφέρει, etc. – are well attested in similar terminological contexts in a wide variety of Galen's works, for example, *On the Causes of Symptoms*, *Introduction to Logic*, *On Differences between Fevers*, *On the Opinions of Plato and Hippocrates*, *On Affected Places*, *On the Method of Healing*, *On the Mixtures and Faculties of Simple Drugs*, *On the Faculties of Foods*, and *Hygiene*.[39]

A well-known stylistic habit of Galen's is likewise reflected in Niccolò's use of *sunt autem alii qui* ... or *sicut et alii*, etc., to mark a transition from one division of medicine to another, particularly in doxographic contexts, for example: 'audire namque est ... et aliquos in curatiuam et uocatam sanatiuam (*sc.* totam artem incidentes); alios autem, qui deposcunt et precustoditiuam his adici mox in prima diuisione, sicut et alii resumptiuam; alii autem utrasque has et cum istis euecticam; sunt autem alii, qui et cirocomicam siue senum educatiuam sicut propriam quandam partem artis adiciunt predictis.[40] The corresponding Greek phrases – εἰσὶ δ'οἵ, ὥσπερ γε καί, καθάπερ γε καί, καθάπερ ἄλλοι τινές, etc. – are well attested in Galen's authentic works.

The author of *PAM* pays extensive attention to the views of medical precursors, notably to the doctrinal disagreements between Empiricists and so-called 'dogmatists.' This is, of course, again characteristic of many of Galen's genuine treatises. Perhaps more significantly, several striking doxographic details in *PAM* concern issues in which Galen showed a keen interest in other works. *PAM* reports, for example, that the third-century BC Empiricist Serapion believed that the third constitutive part of the Empiricists' methodological tripod, namely ἡ τοῦ ὁμοίου μετάβασις (*similis transitio* in the Latin translation), does not have as much value as its other two constitutive parts, i.e. αὐτοψία (*per se inspectio,* Niccolò) and

37 *PAM* 5.3-6.2 (pp. 36-39, 123-24).

38 *PAM* 4.2 (p. 122.18-20).

39 Eg. *De causis symptomatum* 1.5 (VII 108 K); *De differentiis febrium* 1.2 (VII 278 K); *De placitis Hippocratis et Platonis* 2.3.12, 2.5.33, 2.5.81, 5.7.27-28, 6.1.24-25, 6.1.27, 8.1.1 (*CMG* V 4,1,2, pp. 112, 134, 144, 342, 366, 480); *De locis affectis* 1.1, 1.6, 3.3 (VIII 5-6, 48, 141 K); *De methodo medendi* 1.5, 1.6-7, 2.3, 13.1 (X 43, 47, 49-50, 91, 875 K); *De simplicium medicamentorum temperamentis ac facultatibus* 9.1.4 (XII 189-90 K); *De alimentorum facultatibus* 3.23 (*CMG* V 4,2, p. 361); *De sanitate tuenda* 1.5.44-45 (*CMG* V 4,2, p. 12). See also *De pulsuum differentiis* 1.6, 4.5 (VIII 511-12, 732 K); *Synopsis librorum suorum de pulsibus* 9 (IX 458 K); *De antidotis* 1.9 (XIV 49 K); *In Hippocratis De victu acutorum comm.* 2.51 (*CMG* V 9,1, p. 213.13-19); *De ossibus ad tirones*, prooem., 7 (II 735, 736, 756 K); *De difficultate respirationis* 1.2 (VII 758 K); *De anatomicis administrationibus* 4.3, 5.2, 6.2, 9.2 (227, 281, 349, 559 Garofalo); *De neruorum dissectione* 5 (II 835 K); *Aduersus Iulianum* 5.6 (*CMG* V 10,3, p.49); *In Hippocratis Prognosticum comm.* 2.6 (*CMG* V 9,2, p. 266.22-24); *In Hippocratis Epidemiarum I. comm.* 2.49 (*CMG* V 10,1, p. 74.28-31); *In Hippocratis Epidemiarum III. comm.* 1.4, 1.8 (*CMG* V 10,2,1, pp. 18.21-22, 35.7-8); *In Hippocratis Aphorismos comm.* 3.14, 7.58 (XVIIB 595, XVIIIA 170 K); *De constitutione artis medicae ad Patrophilum* 7.4-5, 14.20 (*CMG* V 1,3, pp. 74, 102); *Institutio logica* 4.5 (pp. 10-11 Kalbfleisch).

40 *PAM* 1.2-3 (p. 119.4-12).

ἱστορία, (*hystorie*).[41] In his *Outline of Empiricism* Galen makes a similar point, claiming that it was disputed in antiquity whether Serapion included 'transition from the similar' at all among the constitutive parts of medicine.[42]

More generally, a strong interest in philosophy, especially in Plato, hardly sets this treatise apart from the genuine Galen. Indeed, in 1833, ten years after his attack on the authenticity of the treatise, Kühn (perhaps prompted by afterthoughts about his initial conclusions) drew attention to the fact that the author of *On the Parts of the Medical Techne* approvingly referred to Plato's *Sophist* and *Statesman*.[43] The context of *PAM*'s reference to these two Platonic dialogues is the question of the correct method of division, and it is precisely in this context that Galen invoked the same Platonic dialogues in several of his treatises of undis-puted authenticity, for example, in *On the Method of Healing, On the Opinions of Hippocrates and Plato, On Differences between Pulses*, and *Against Lycus*.[44]

Even more significantly, the age-old issue of the correct division of the medical *techne* into its parts, the history of such divisions, the nature of the relations between the parts, and the question whether every specialized form of medical expertise is a distinct *techne*, were often on Galen's mind in his works of undisputed authenticity. While *On the Parts of the Medical Techne* offers the most sustained Galenic treatment of these topics, they are also addressed, for example, in Galen's *Hygiene, Thrasybulus, Outline of Empiricism, On the Composition of Drugs According to Places, On the Composition of Drugs According to Kinds*,[45] and in a work most but not all modern scholars regard as authentic, *Medical Techne* (*Ars medica*).[46] While *On the Parts* in many respects goes considerably beyond what Galen said elsewhere about the division of the *techne,* this is to be expected in a treatise devoted exclusively to the subject.

41 *PAM* 6.6 (pp.125.21-26; cf. pp. 41.35-43.2).

42 Gal., *Subfiguratio emperica* 3-4 and 9 (pp. 49.4-50.25, 69.29-74.23 Deichgräber); see n. 88 below.

43 *PAM* 9.5 (pp. 48-49, 129); see Kühn, *Galeni opera* (n. 1 above), vol. XX, p. xiv.

44 *Methodus medendi* 1.3 (X 26.1-7 K); *PHP* 9.5.11-13 (*CMG* V 4,1,2, p. 566.10-26 De Lacy); *De pulsuum differentiis* 4.7 (VIII 736.1-5 K); *Adversus Lycum* 3.15 (*CMG* V 10,3, p. 12.11-14 Wenkebach).

45 Eg. *De sanitate tuenda* 1.1-5 (*CMG* V 4,2, p. 3 Koch); *Thrasybulus*, passim, eg. chapters 2, 5-6, 8-9, 15, 23-28, 30-31, 33, 35, 39-42, 47 (*Scr. min.* III 33-100 Helmreich); *Subfiguratio empirica* 5-6 (pp. 51-61 Deichgräber); *De compositione medicamentorum secundum locos* 1.2 (XII 434 K); *De compositione medicamentorum per genera* 3.2 (XIII 604 K). For further evidence of the extensive Greek debates about the correct division of medicine into its parts see ps.-Gal., *Introductio siue medicus* 5 (XIV 684 K = Erasistratus, fr. 32 Garofalo) and 7-8 (XIV 689-95 K); ps.-Gal., *Definitiones medicae* 10-11 (XIX 351-52 K); Celsus, *Medicina*, prooem. 9 (*CML* I, p. 18 Marx); *Anonymus Bruxellensis,* ch. 40, in Max Wellmann, *Fragmentsammlung der griechischen Ärzte: Die Fragmente der sikelischen Ärzte Akron, Philistion und des Diokles von Karystos* (Berlin, 1901) 233; ps.-Soranus, *Quaestiones medicinales* 12-15, 23-24, 28, 30, in V. Rose, *Anecdota Graeca et Graecolatina*, vol. 2 (Berlin, 1870) 249-251; Sextus Empiricus, *Math.* 1.95. For further evidence of ancient divisions of medicine see Ch. Daremberg, *Hippocrate* (Paris, 1843), 387 ff.; Schöne, *Galenus* (n. 8 above) 17 n. 1; Ludwig Englert, *Untersuchungen zu Galens Schrift Thrasybulos*, Studien zur Geschichte der Medizin 18 (Leipzig, 1929) 21-31; H. von Staden, *Herophilus. The Art of Medicine in Early Alexandria* (Cambridge, 1989; repr. 1994) 89-112. See also n. 72 below.

46 The authenticity of *Ars medica,* long taken for granted, has been questioned by J. Kollesch, 'Anschauungen von den APXAI in der Ars medica und die Seelenlehre Galens,' in *Le opere psicologiche di Galeno*, ed. by P. Manuli and M. Vegetti (Naples 1988) 215-229, and by L. García Ballester, 'On the origin of the "six non-natural things" in Galen', in *Galen und das hellenistische Erbe*, ed. by J. Kollesch and D. Nickel, Sudhoffs Archiv Beiheft 32 (Stuttgart 1993) 105-115. More recently its authenticity has been vigorously defended by V. Boudon, 'L'*Ars medica* de Galien un traité authentique?', *REG* 109 (1996) 111-56, and again in her edition of *Ars medica* in the Budé series: *Galien*, tome II (Paris 2000) 157-64.

Further affinities between *PAM* and Galen's genuine works will be pointed out in Parts 4 and 5 (below). If there are discrepancies between this treatise and the corpus of Galen's authentic works, they are no greater than similar inconsistencies and discrepancies both within some of his treatises and between some of his works. Some of the discrepancies are a result of Galen's overtly expressed uncertainty about a view he had previously presented as certain. But the numerous stylistic, doctrinal, and nomenclative resemblances between *PAM* and his other works, along with the ample cross-references to authentic Galenic works, cumulatively render it plausible that Galen was the author of *On the Parts of the Medical Techne*. In the pages that follow I therefore shall refer to the writer as Galen, although incontrovertible certainty about the authorship is unattainable, notably in the absence of the Greek original.

4. Division and dissection: anatomy as methodological model

More than thirty different divisions of the medical *techne* into its parts are introduced in *PAM*. Some are presented as superseded historical relics, some as honourable precursors, and others as new theoretical possibilities. Already in the first chapter, devoted largely to a critical historical survey of non-Empiricist divisions, Galen introduced fifteen alternative divisions, none ascribed to a specific author. In complexity they range from a simple division into two branches (therapeutics and hygiene) or three (the famous triad pharmacology, surgery, and regimen[47]) to a sometimes less than lucid identification of up to thirteen additional parts: the prophylactic, convalescent, euectic (pertaining to the maintenance of a 'good condition', εὐεξία), geriatric, paediatric, cosmetic, beautificatory, vocal-rectificatory, vocal-natural, physiological, pathological, semiotic, and aetiological parts of the *techne*. In the second chapter Galen turned to disagreements within the Empiricist school about the proper division of the medical *techne* into its final and constitutive parts,[48] and in subsequent chapters yet further divisions – as well as distinctions between different levels of magnitude to which different divisions belong – are introduced.

The multitude of previously proposed divisions as well as the many disagreements among their proponents are depicted as having led to the perplexity of Galen's addressee ('my very

47 A *locus classicus* for this tripartite division is Aulus Cornelius Celsus, *Medicina*, prooem. 9 (*CML* I, p. 18 Marx = p. 16 Mudry = p. 4 Serbat). The context in Celsus suggests that the division dates to the late fourth or to the third century BC. See also n. 125 below (Soranus); Gal., *Subfig. emp.* 5 (p. 52.12-13 Deichgräber); *PAM* 6.1-4 (pp. 38-41, 124-25); Anonymus Bruxellensis 40 (p. 233 Wellmann). See von Staden, *Herophilus* (n. 45 above) 99-100. For the controversial claim that the early 'Pythagoreans were among those who accepted this particular classification of medicine' and that 'the sequence of the various parts of the healing art in the Pythagorean doctrine is the same as in the Hippocratic Oath, dietetics coming first, pharmacology next, surgery last' see L. Edelstein, *Ancient Medicine* (Baltimore 1967) 21-31. Edelstein, who relied heavily on Iamb. *VP* 29.163 (p. 92.4-17 Deubner/Klein), also argued that the 'Pythagorean' hierarchical valorization reported by Iamblichus (regimen being the most important, followed by pharmacology, with surgery ranked last) or by his source Aristoxenus influenced the Hippocratic Oath. Cf. Hp. *Aph.* 7.87.

48 *PAM* 2.1 (pp. 26-27, 120.7-15). See also K. Deichgräber, *Die griechische Empirikerschule*. 2nd ed. (Berlin/Zurich 1965) 119-130 (fr. 39-64), 289-305.

dear Justus'),[49] which in turn supposedly led Galen to compose the work.[50] The competing divisions not only afforded Galen an opportunity to display his mastery of doxography (chapters 1-2) to Justus and to other members of his audience,[51] but also to develop and justify his own views at length (chapters 3-9). The groundwork for his principal theory – that a correct division of the medical *techne* will end with the discovery of its smallest parts or its most elementary 'theorems' or forms of knowledge (in Niccolò's Latin the *minimae et elementales particulae, paruissimae particulae, indiuisibiles particulae, elementarissimae scientiae, ultimae et minimae et simplices notitiae, ultima et elementalia theoremata*[52]) – is laid by means of an elaborate analogy between the division of the medical *techne* and dissection.

First introduced in chapter 3[53] and then repeatedly invoked in subsequent chapters, the anatomical analogy formally is a response to a question raised in chapter 3.1: what are the smallest parts or branches of the medical *techne*, or more precisely, which of its parts should be called the smallest (probably τὰ ἐλάχιστα μόρια)?[54] Galen's answer begins by stating that, just as in the dissection of the body one tries to learn the parts of the body first through larger parts and then down to its smallest parts κατ' εἶδος,[55] i.e. all the way down to the 'perceptible elements' (τὰ αἰσϑητὰ μόρια), so too one should try to learn the smallest parts of the medical *techne*. The anatomical side of the analogy is developed first (ch. 3.3-6): it starts with a division of the body into its major parts – head, chest, belly, and limbs in the Arabic version, but head, hands, and feet in Niccolò's translation – and then proceeds through various subdivisions. The head, for example, is subdivided into ears, face, nose, and neck, or, by a different division, using 'head' in the more restrictive sense of 'the part of the body on which hair grows', into the brain, meninges (*dura* and *pia mater*), the skull, and the

49 Lyons, *PAM* 1.1 (p. 25.1-2): 'My dear Justus, I think you have good reason to be confused on the subject of the parts of medicine'; cf. Niccolò (p. 119.3-4): '*De partibus medicatiuae, Juste dilectissime, conuenienter mihi uideris dubitare, cum alii et alii aliter eas distinguant*'. Kollesch, Nickel, and Strohmaier (*PAM* 119.4) delete *et alii*. On the possibility that the addressee might be the eye doctor Justus whom Galen mentioned in *De methodo medendi* 14.19 (X 1019K) see Schöne, *Galenus* (n. 8 above) 21-22. Another possibility is the husband of the patient at *De praecognitione* 5.6 (*XIV* 626 K = *CMG* V 8, 1, p. 94. 22 Nutton); see Nutton, *ad loc.* (pp. 186-87).

50 Galen leaves this impression with his readers (see n. 49 above). Indeed, the summary of ch. 1 (omitted from the *CMG* edition – as are all the summaries inserted before each of the subsequent chapters – but included by Schöne) says that this 'little book was composed in response to Justus' request': 'Incipit libellus Galieni de partibus artis medicatiuae factus ad petitionem Iusti, qui erat in mente perturbatus pro eo, quod non concorditer sed diuerse medici diuiserunt ipsam medicatiuam'; Schöne, *Galenus* (n. 8 above) 23.1-3.

51 Although formally addressed to Justus, the treatise sometimes addresses the audience in the plural, eg. *PAM* 4.5 (122.35), 9.1 (128.22): *sciatis*, in Niccolò's rendering. But at the end Galen again refers to his reader in the singular; 9.6 (p. 129.13-15): *te, considera*. See also 6.6 (125.26-27): *...tu eum qui per indicationem et methodum ingressum ad talia nosti ...*, and the discussion by Schöne, *Galenus* (n. 8 above) 20.

52 For these and related expressions see eg. *PAM* 3.1-4, 4.2, 8.3, 9.4 (pp. 121, 122, 127, 128, 129). The corresponding Greek expressions perhaps were τὰ ἐλάχιστα καὶ στοιχειώδη μόρια, αἱ στοιχειωδέσταται ἐπιστῆμαι, and τὰ στοιχειώδη καὶ ἔσχατα ϑεωρήματα. Galen's use of most of these expressions – albeit sometimes in different contexts – is independently attested. For τὰ στοιχειώδη μόρια see eg. *De naturalibus facultatibus* 1.6, 1.7 (*Scr. min.* III, pp. 110.24, 112.16-23), *De sanitate tuenda* 6.2.1 (*CMG* V 4,2, p. 169.10-11). For τὰ στοιχειώδη ϑεωρήματα see *De animi cuiuslibet peccatorum dignotione et curatione* 5.16 (*CMG* V 4,1,1, p. 59.2-3). See n. 108 below.

53 *PAM* 3.2 (pp. 28-31, 121.9-11).

54 *PAM* 3.1 (pp. 28-29, 121.4-9).

55 *PAM* 3.2 (p. 121.10): *secundum speciem* (Niccolò); Lyons (29 n. 2) observes that Ḥunain (or Isḥāq?) mistranslated εἶδος by 'ṣūra'.

pericranial arteries, veins, membranes, nerves, and skin.[56] Some of these in turn are subdivided further until one reaches what Galen calls the smallest 'perceptible elements'.[57]

The relation of these perceptible elements (τὰ αἰσθητὰ στοιχεῖα), first, to the four classical elements (earth, water, air, fire) and, secondly, to the 'instrumental parts' is not adequately clarified in this treatise. But from a number of other works in which Galen likewise introduced the notion of 'perceptible elements' it is clear that, while he used 'element' (στοιχεῖον) in a variety of ways, he consistently identified the smallest *perceptible elements* as the homoeomerous parts (such as fibres, membranes, flesh, fat, bone, cartilage, ligaments, marrow and nerves), which, he said, are simple (ἁπλᾶ) and uniform.[58] Fire, air, water, and earth, by contrast, while also 'simple elements' (ie. not composite, οὐ συνθετά), are the 'first elements *by nature*' (τὰ πρῶτα στοιχεῖα φύσει) or 'the real elements' (τὰ ὄντως στοιχεῖα).[59] Furthermore, *qua* elements of compound bodies, these 'first natural elements' are imperceptible, whereas the homoeomerous elements are accessible to sensory perception. What distinguishes the smallest 'perceptible elements' or parts from the instrumental parts (ὀργανικὰ μόρια or ὄργανα, such as brain, heart, liver, lung, eye, hand, finger, leg, and tongue) is not only that each instrumental part is composite in a double sense – it is composed both of the four elements that are first 'by nature' and, at a different level, of some first 'perceptible elements' – but especially that each instrumental part is responsible for an activity that is necessary or useful for the living being as a whole.

Through a series of anatomical subdivisions Galen arrived (3.4) at an example of the 'perceptible elements' or 'smallest parts' of the body, namely, the fine structure of the transverse and lengthwise fibres that constitute the two tunics or coats of the arteries. The procedure Galen envisioned on the bodily side of the analogy is therefore reasonably clear: dissection proceeds not only by moving from larger to increasingly smaller pieces of anatomical territory but also by determining functional differentiae and by distinguishing between functional levels. He also emphasized that the smallest 'perceptible elements' at which a successful dissection finally arrives can be quite varied in nature (inasmuch as the homoeomerous parts include fat, bone, marrow, fibre, etc.), even if each such element is uniform and anatomically indivisible.

The *techne* side of the analogy between dissection and the division of the *techne* is elaborated sporadically throughout the rest of the treatise. Galen does not seem to have doubted the usefulness of the anatomical 'cuts' or divisions into instrumental parts and then into their perceptible elements as a model for the division of the medical *techne* (even if he

56 *PAM* 3.3 (pp. 30-31, 121.11-20).

57 *PAM* 3.3 (pp. 30-31). Lyons (*PAM*, p. 31.15-16) translates the final phrase in 3.3 as follows: '... and to show there how the body can be divided until the sensible elements are reached'. Niccolò here omits 'sensible' or 'perceptible': '...*quamcumque ostendere incisionem corporis usque ad elementa*'; but he preserves *elementa sensibilia* in 3.2, 3.5, and 4.1 (pp. 121.10, 121.28, 122.1, 122.9). See n. 61 below.

58 On 'perceptible elements' and their identification with homoeomerous parts see eg. Gal., *De elementis* 1.1-7, 6.28-29, 8.11 (*CMG* V 1,2, pp. 56, 110, 126); *De naturalibus facultatibus* 1.6, 3.15 (*Scr. min.* III, pp. 109.18-19, 110.9-15, 255.19-20); *De sanitate tuenda* 6.9.2 (*CMG* V 4,2, p. 184.20-26); *In Hippocratis De articulis comm.* 3.81 (XVIIIA 597.3-7 K). At times Galen also refers to the perceptible elements as τὰ φαινόμενα στοιχεῖα, eg. *De elementis* 1.7, 6.28 (*CMG* V 1,2, pp. 56.17, 110.15-16).

59 Eg. Gal., *De elementis* 1.1-9, 6.28 (*CMG* V 1, 2, pp. 56-58, 110); see also *In Hippocratis De natura hominis comm.* 1.44 (*CMG* V 9, 1, p. 53.25-27).

acknowledged the validity of other approaches to identifying branches of the *techne*, as will be shown below). The *techne* of couching a cataract, for example, is said to be a part of the medical *techne* as a whole in the same sense that an instrumental part, such as the eye, is a part of the human body as a whole.[60] Yet the couching *techne* in turn is composed of other *technai* that are its smallest elements or its 'minimal sciences', just as, says Galen, an instrumental part of the body is composed of perceptible elements or 'smallest and most elemental parts'.[61] He enumerates several such 'elemental' *technai* or 'minimal sciences' that constitute the 'instrumental' *techne* of couching. They include knowledge of the entire form of the eye, the *techne* of determining the exact part of the eye where one should couch, the *techne* or 'minimal science' of evaluating the disposition of the eye in a given patient at a given time, the *techne* of breaking up and shifting the cataract, and that of removing it from the eye.[62] These are examples of 'the simple and smallest elements' of the *techne*: just like the 'perceptible elements' of the body, none of them is susceptible to further cutting or division into smaller parts of the *techne*.

Already towards the end of this first, paradigmatic elaboration of the analogy between anatomical dissection and the division of medicine, however, Galen showed unease with its results: 'But the whole treatment of cataracts requires some other forms of knowledge'.[63] He identified these 'other forms' as being twofold κατὰ γένος: one is the preparative science (probably τὸ παρασκευαστικὸν μόριον/μέρος or ἡ παρασκευαστικὴ τέχνη),[64] the other the science of completion through after-care (probably ἀποϑεραπευτικόν).[65] Exactly how these two branches of the *techne* fit into Galen's division of medicine into its 'instrumental' branches and its 'elemental parts' or 'smallest sciences' is not clear. His apparent use of κατὰ γένος (*secundum genus* in Niccolò's translation)[66] might signal that the preparatory *techne* and the one devoted to completing the treatment by means of after-care do not belong among the smallest or elemental parts of the *techne*, inasmuch as both skilled preparatory treatment and expert after-care are required in the therapy of numerous diseases besides

60 *PAM* 4.1 (pp. 32-33, 122.6-8).

61 *PAM* 4.1 (32-33, 122.8-11): '... each of the parts of cataract couching is the equivalent of a sensible element of the body' (Lyons); *singula aut<em> eius* (scil. *apunctiuae artis*) *particularum proportionalis est sensibilibus corporis elementis* (Niccolò). Niccolò's *proportionalis* probably translates ἀνάλογον, as Schöne (*ad loc.*) suggested.

62 *PAM* 4.2 (pp. 32-33, 122.11-20).

63 *PAM* 4.3 (pp. 32-33, 122.22-23). The Arabic and Latin versions are not in full agreement. Cf. Lyons (p. 33.30-31): 'but for the full cure of the fluid generated in the eye you need two more [scil. branches of knowledge] in addition to them'; and Niccolò (p. 122.22-23): 'uniuersa aut<em> cura suffusionum indiget aliquibus aliis noticiis secundum genus'. If Niccolò's *secundum genus* (missing from the Arabic) accurately reflects the Greek original (probably κατὰ γένος), it would be consistent with Galen's apparent use of (κατὰ) γένος elsewhere in this treatise, eg. *PAM* 2.1, 7.1, 8.3, 8.5 (pp. 26-27, 44-45, 120.8, 126.6, 127.25, 128.5). See n. 76 below.

64 *PAM* 4.3 (pp. 32-33, 122.24). Schöne, *Galenus* (n. 8 above) 28 n. on ll. 132-33, followed by Lyons (184.1), suggested προπαρασκευαστική, which is, however, not attested in Galen's works, whereas παρασκευαστικός is (usually qualifying γυμνάσιον or τρίψις); see eg. *De sanitate tuenda* 2.4.52, 2.6.5, 2.6.7, 2.7.19, 3.11.2-4, 3.11.10-11, 3.12.1 (*CMG* V 4, 2, pp. 52, 55, 57, 97, 98, 99); *De tumoribus praeter naturam* 3 (VII 716.12 K).

65 Lyons, *PAM* 4.3 (p. 33.33): 'the other provides the completion of the cure' (Ḥunain); Niccolò (p. 122.24): *secunda autem nominata apotherapeutica.* ἀποϑεραπευτικός is independently attested as a Galenic technical term; see eg. *De sanitate tuenda* 2.6.6, 3.2.54, 3.6.11, 3.7.13, 3.11.2-4, 3.11.10, 3.12.1-6, 4.4.13, 5.10.30, 6.7.7 (*CMG* V 4, 2, pp. 55, 79, 87, 89, 97, 98, 99, 157, 181).

66 *PAM* 4.3 (pp. 32-33, 122.23). The Arabic version omits κατὰ γένος; see nn. 63 above and 76 below.

cataracts. As more general branches of medicine, τὸ παρασκευαστικόν and τὸ ἀποθεραπευτικόν therefore cut across many different disease-specific divisions of expertise. In their disease-specific applications – for example, expertise in preparing the eye for a cataract operation, or post-operative expertise in the treatment of a couched eye – the preparatory and post-operative *technai*, however, become 'elemental parts' of the *techne* of couching. It is to these disease-specific preparatory and post-operative branches of medicine that Galen appears to refer in chapter 4 as being 'amongst the smallest ultimate little parts of medicine'.[67]

Galen was aware that the model of division inspired by dissection presents further difficulties, and he tried to address some of these in order to sustain the model. He raised the question, for example, of where in such a division of medicine one may locate the *techne* (a) of treating inflammation (as a generically definable condition), and (b) of treating inflammation of a particular part. The medical *techne*, he said, consists of treatments of diseases, and the knowledge of treatment is divided into its 'primary and fundamental' or 'most appropriate' parts. These are, first, the forms of expertise that deal with the treatment 'of the individual and simple diseases, next of the compound diseases, and next of those that affect certain parts of the body, first simple and then compound parts.'[68] Inflammation is a disease, but, said Galen (in explicit opposition to the Empiricists' view), there is no single *techne* for treating inflammation, understood generally or absolutely (ἁπλῶς, *simpliciter*), as little as there is a single general *techne* for treating a fracture or an ulceration or a painful reddish swelling of the skin (*erysipelas*[69]), regardless of where it might occur. Just as 'instrumental part' , as a general conceptual category, may be said to be a part of the body, so 'the treatment of inflammation' is a part of medicine. But more relevant to the success of medicine is knowledge of how to treat inflammation of specific parts, for example, inflammation of the external tunic of the eye (ophthalmia), inflammation of the lung (pneumonia), inflammation 'of the tunic encircling the ribs' (pleuritis), and so on. Each of these forms of expertise is like a specific 'instrumental part' of medicine, and each therefore requires further subdivision into its elemental parts or 'smallest sciences'. Galen insisted that the *techne* of treating the *genus* 'inflammation' (Lyons: 'used generically') differs from the *technai* of treating the various species of inflammation, 'just as the generic term "organ" differs from a specified organ',[70] and he referred his audience to an earlier demonstration – in his *Method of Healing*[71] – of such differences.

67 Lyons, *PAM* 4.4 (p. 35.5-6): 'the elements of the art of medicine'; cf. Niccolò (p. 122.34): *minime et ultime particule*.

68 Lyons, *PAM* 5.2 (p. 37.9-11). Niccolò's rendering (123.20-22), however, does not include the reference to diseases of body parts: *cum ars medicatiua existat de curis egritudinum, ad primas et principalissimas partes diuiditur notitie cure simplicium egritudinem, deinde compositarum*. In both the Arabic and the Latin translations this passage is followed by a cross-reference to Galen's discussion of simple vs. compound diseases in *De morborum differentiis*.

69 On the ancient meanings of *erysipelas* see M. D. Grmek, *Diseases in the Ancient Greek World* (Baltimore/London, 1989) 129.

70 Lyons, *PAM* 4.6 (p. 35.34); Niccolò (123.13-14): *sicut organum quod generaliter intelligitur, aliud est a determinato organo*.

71 *PAM* 5.6 (34-35, 123.15-16). He probably has in mind Book 13 of *Methodus medendi* (X, 874-944 K), where Galen identifies the generic treatment of inflammation as, in principle, consisting of reducing excessive heat and reducing or expelling an excess of blood, and where he repeatedly (especially in 13.7 ff.; X, 897 ff. K) distinguishes between

Eager though Galen was to sustain the analogy between anatomical dissection and the division of the medical *techne,* the need to generate other parts of the *techne* and to accommodate more complex perspectives appears to have prompted him to move increasingly away from the anatomical analogy without, however, ever abandoning it completely. In chapter 5 Galen made a transition to a division of the medical *techne* based on three fundamental bodily 'conditions' or 'dispositions' (probably διαθέσεις in the original): a diseased disposition, a healthy disposition, and a disposition which is 'neither' healthy nor diseased (οὐδέτερον). In this third disposition the body is neither in perfect health nor impeded in its habitual activities by its imperfect health. The tripartition presented here has a striking parallel in the definition of medicine in *Ars medica*: ἰατρική ἐστιν ἐπιστήμη ὑγιεινῶν καὶ νοσωδῶν καὶ οὐδετέρων.[72] In *Ars medica*, the triad is presented as part of a definition of medicine, whereas here it functions as a basis for the division of medicine into its parts. Yet in both treatises Galen used the threefold structure to generate divisions that identify parts of medicine, and in both the tripartition is mapped onto the concept 'disposition' (διάθεσις),[73] even if in *Ars medica*, unlike *PAM*, 'disposition' in turn is, in the first instance, mapped onto the threefold distinction between body, cause, and sign.[74]

It should be noted, however, that in *On the Parts of the Medical Techne* Galen expresses himself more tentatively than in *Ars Medica* about the division into three dispositions:

> *If* in fact there is another, third disposition, which *they* call 'neither' [healthy nor diseased], it is necessary that a different part of the *techne* provide for it. It *seems*, however, that the disposition of those who clearly suffer from *dyskrasia* – i.e. of those still capable of performing their customary activities – is of such a kind but not of those who are ill.[75]

In a subsequent chapter Galen again signalled his uncertainty about the validity of the threefold division, commenting that the subject matter which most properly belongs to medicine is 'the dispositions of the body, *whether there are two, three or more* classes of

general expertise in reducing such excesses in homoeomerous parts and specific expertise in achieving these goals in particular 'instrumental parts' (ὀργανικὰ μέρη). See also *Meth. med.* 13.11 (X, 901 K) on knowing the universals (τὰ καθόλου) vs. being practised περὶ τὰ κατὰ μέρος.

72 *Ars medica* 1b.1 (I, 307 K = 276 Boudon). Galen's elaboration of this definition provides the basic structure for *Ars medica* 1b.1-36.8 (I, 302-405 K = 276-386 Boudon), ie. for most of the rest of the treatise. The elaboration begins by specifing that each of the three – healthy, diseased, and 'neither' – can be predicated in three ways, namely as body, as cause, and as sign. The 'healthy', for example, can refer to the *bodies* that are recipients of health, to the *causes* that produce and preserve health, and to the *signs* that indicate health. Of this elaboration there is little or no trace in *PAM.* On the Herophilean origins of the threefold division into healthy, diseased, and 'neither' see H. Schöne, *De Aristoxeni Περὶ τῆς 'Ηροφίλου αἱρέσεως libro tertio decimo a Galeno adhibito* (Bonn, 1893) 24-26; von Staden, *Herophilus* (n. 45 above) 103-08; V. Boudon, 'Les définitions tripartites de la médecine chez Galien,' *Aufstieg und Niedergang der römischen Welt* II: 37.2 (Berlin, 1994) 1468-90; *ead.* (ed., trans.), *Galien, tome II: Exhortation à l'étude de la Médecine, Art médical* (Paris, 2000) 178-92, 396-98. See n. 104 below.

73 Eg. *PAM* 5.4, 6.1 (pp. 36-39, 123.30, 123.32, 124.21): 'condition', and 'state' (Lyons), *dispositio* (Niccolò); *Ars medica* 3.1-4, 36.8-37.1 (I, 313, 405-06 K = 281-82, 386 Boudon).

74 See n. 72 above.

75 *PAM* 5.4 (36-37, 123.29-124.1). Niccolò's literal rendering, on which my translation is based, here seems closer to the original than does the Arabic version.

these,'[76] and shortly afterwards he abandoned the tripartite division for a bipartite dispositional model, as will be shown below. The provisionality and vacillation expressed in some of these passages might indicate that the treatise is one of Galen's later works. As Vivian Nutton has pointed out, in another late work, *On my Own Opinions,* Galen also displays greater doctrinal reticence, distinguishing more carefully than he had earlier in his career between what he knows to be certain, what is plausible but unproven, and what he remains in doubt about.[77]

Despite the tentative tone of the passages quoted above, Galen boldly proceeded to use the three 'dispositions' not only to generate new parts of medicine but also to reposition branches of medicine that he had already identified. The parts of medicine previously identified and classified by means of the analogy with dissection (for example, the *techne* of couching and all its subdivisions, the *techne* of treating inflammation and its parts) now are subsumed under therapeutics (probably τὸ θεραπευτικὸν μέρος τῆς τέχνης in the original). The therapeutic *techne* in turn is assigned to the diseased disposition, while the two other major branches of medicine – 'hygiene' and a nameless 'neutral' one – are correlated with 'healthy' and 'neutral' dispositions, respectively.

The disposition called 'neither' in turn is used to generate several smaller parts of medicine, such as the prophylactic (προφυλακτικόν, *precustoditiua* Niccolò), recuperative (ἀναληπτικόν, *resumptiua*), and geriatric parts (γηροκομικόν, *cirocomica*).[78] Without expressing disagreement or agreement, Galen reported that 'some people' also located the paediatric part of medicine within the larger branch devoted to bodies that are neither healthy nor ill.[79] Galen himself apparently also regarded the part of medicine that aims at maintaining a sound condition (εὐεξία) as belonging to the larger branch devoted to the 'neutral' disposition, adding that some call it the 'euectic' part (τὸ εὐεκτικὸν μέρος, *euectica pars*), others 'the art of exercise' or 'gymnastics' (Lyons: 'the art of massage').[80]

On the basis of the three dispositions Galen concludes that:

> the primary and main branches, then, of the *techne* that concerns itself with the body are these: the therapeutic, the recuperative, the prophylactic, the geriatric, the paediatric, together with those that are concerned with the preservation of health and of good

76 *PAM* 7.1 (42-43, 126.6). The Latin *secundum genus* (κατὰ γένος?) here again is omitted in the Arabic (see n. 63 above). Subsequent sections of *PAM*, explicitly harking back to a view Galen had espoused in his *Thrasybulus*, abandon the tripartite division for a twofold version: 'The primary difference in dispositions is double, ... and it is our custom to call one of these health but the other diseases'; see *PAM* 7.2 (pp. 42-43, 126.6-10), 8.1-2 (pp. 44-45, 127.15-16).

77 Eg. *De propriis placitis* 1.4, 11.2, 13.7, 14.4-15.2 (*CMG* V 3,2, 56.7-11, 90.26-92.3, 108.11-110.3, 114.6-118.5 Nutton).

78 *PAM* 5.5-5.6 (36-37, 124.4-11).

79 *PAM* 5.7 (36-39, 124.11-13). Schöne, *Galenus* (n. 8 above) 23 (n.on line 17) and 31 (n.on line 198) suggests that παιδοτροφικόν (*scil.*μέρος τῆς ἰατρικῆς) was Galen's original term. Niccolò's *pedotrofica* lends strong support to Schöne's suggestion, but the fact that παιδοτροφικόν is not attested as a Galenic term (indeed, I am not aware of its use by any ancient Greek author) gives one pause.

80 *PAM* 5.7 (38-39, 124.14-16). The Arabic translation presents this as Galen's own view, but Niccolò's version attributes it – as well as all the nomenclative alternatives – to 'others': 'adhuc autem alii addunt aliam partem, que est circa euexiam, que uocatur ab aliquibus quidam euectica, ab aliquibus autem exercitastica et ginastica, ut sint prime et maxime partes artis que est circa corpus ...' The corresponding Greek terms may have been εὐεκτικόν and παιδοτριβικὸν ἢ γυμναστικόν. See also *PAM* 1.2, 7.3 (pp. 24-25, 42-43, 119.10, 126.15) on *euectica* (*pars*).

condition (*euexia*). These branches may all be subdivided, as we showed earlier in our division of the therapeutic part,[81] and this leads to the types of simple and indivisible sciences of which we said that they are the 'elements' of medicine.[82]

Despite Galen's expressions of uncertainty about the existence of the third or 'neutral' disposition, he described no less than five of the seven 'primary and main branches' (τὰ πρῶτα καὶ μέγιστα μέρη τῆς τέχνης; *prime et maxime partes artis*, Niccolò) enumerated in this passage as devoted to the bodily disposition that is 'neither' healthy nor diseased. By contrast, the healthy disposition, like the diseased disposition, here is assigned only one of the seven 'first and greatest' branches of medicine, viz. hygiene and therapeutics, respectively (although Galen was quick to underscore that hygiene and therapeutics are themselves divisible down to their 'elemental' parts). Such asymmetries in division were, quite sensibly, of no concern to Galen; not only did he view some main branches of medicine as composed of more indivisible 'elemental' parts than others, but he also did not believe that each of those bodily dispositions generated the same number of 'primary and main branches' of medicine. Furthermore, his aim in *PAM* does not seem to have been to provide an exhaustive catalogue of all parts of medicine, but rather to explore all the valid ways of arriving at the smallest parts – the 'perceptible elements' – of the medical *techne*, ie. at the knowledge of the 'elements' that is essential to a true mastery and an efficacious practice of the *techne*.

Even after this enumeration of the major parts of medicine Galen recognized, however, that further perspectives were required, if he were to account adequately for certain widely recognized branches of medicine. In his own divisions, he had not yet explicitly included surgery, pharmacology, regimen, anatomy, physiology, and semiotics (prognosis, diagnosis), nor had he refuted their legitimacy as parts of the *techne* (although several of these branches had been mentioned in his opening doxography). So, having previously identified and classified various parts of the medical *techne* by deploying the analogy with dissection, the distinction between complex and simple diseases, the distinction between generic and specific or local treatment of a disease, and the three dispositions, he now turned (chapter 6) to divisions that have other foundations. In doing so, he also reverted to a doxographic mode, in part accepting precursors' views, in part criticising them.

The first of the new divisions is one based on differences in 'activities' or 'operations' in medicine (κατὰ τὰς τῶν ἐνεργειῶν διαφοράς is Schöne's plausible reconstruction; *secundum operationum differentias,* Niccolò [83]). Some, he said, claimed this as the basis for the famous division of medicine into surgery, pharmacology, and regimen as the primary and most important branches,[84] while others expanded the notion of 'activities' to add semiotics

81 Galen here refers back to the examples of couching a cataract and treating inflammation; *PAM* 4.1-6 (pp. 32-35, 122-123). But see also *PAM* 6.3 (pp. 40-41, 125.3-8) on the Empiricists' division of therapeutics into regimen, pharmacology, and surgery; see n. 88 below.

82 *PAM* 5.8 (pp. 38-39, 124.16-20). Niccolò's translation reverses the order in which the hygienic and euectic parts are listed (*euectica et sanatiua*) but otherwise is in agreement with the Arabic version.

83 Schöne, *Galenus* (n. 8 above) 31 n. on line 208. For Galen's uses of αἱ διαφοραὶ τῶν ἐνεργειῶν see eg. *Thrasybulus* 23 (*Scr. min.* III 61.16), *De symptomatum differentiis* (VIII 55 K), *De methodo medendi* (X 249 K), *De sanitate tuenda* 1.5.9 (*CMG* V 4,2, p. 9.4-6).

84 *PAM* 6.1 (pp. 38-39, 124.21-25).

(or its subdivision into prognostic and pathognomic parts).[85] Still others defined the division into surgery, pharmacology, and regimen as based not on 'activities' or 'operations' but on the material causes (diet, drugs, surgery) of health and illness.[86] Another approach to the same division, says Galen, was based on the entire range of medical knowledge, thus adding physiology, aetiology, and so on, to surgery, pharmacology, and regimen.[87] In this context Galen once again took up the Empiricists' division of medicine into three final parts and two or three constitutive parts, depicting one of the three 'final' parts, therapeutics, as subdivided into surgery, pharmacology, and regime.[88] Galen sharply criticised some features of the Empiricist division, while praising others.

Upon completion of this interlude of critical doxography Galen reaffirmed the centrality of the division according to dispositions, arguing that the subject matter of medicine 'may most properly be taken to be the dispositions of the body.'[89] But, as indicated above, here he drew back from the threefold division and returned to the dispositional duality – a 'healthy' and a 'diseased' disposition – espoused in his *Thrasybulus*. Galen now subdivided each of these two basic dispositions into a habitual and a transient such disposition (probably διάθεσις καθ' ἕξιν vs. διάθεσις κατὰ σχέσιν in the original).[90] Abandoning dispositional tripartition for bipartition, and differentiating enduring dispositions from those that are temporary, obviously entailed reallocating all the parts of medicine previously correlated with the 'neutral' dispositions. According to the Arabic translation, which here appears to offer a more plausible version of the original than does the Latin, Galen endorsed the correlation of hygiene with a 'habitually healthy disposition' and identified the euectic part as one of the subdivisions of hygiene.[91] By contrast, the prophylactic, recuperative, and geriatric parts of

85 *PAM* 6.2 (pp. 38-41, 124.34-125.3). Niccolò's use of *pathognomonicam* (p. 125.2) accurately reflects Galenic usage; for the distinction between προγνωστικός and παθογνωμ(ον)ικός see eg. Gal., *In Hippocratis De officina medici comm.* 1.4 (XV11IB 603 K), *In Hippocratis Prorrheticum I. comm.* 1.4 (*CMG* V 9, 2, p. 19.6-14).

86 Lyons, *PAM* 6.1 (p. 39.19-22): 'It has been thought by some that this kind of division can be based on material causes, in that everything whose function it is to bring illness to the body or to give it health has its origin either in diet, drugs or surgery'. Cf. Niccolò, *PAM* 6.1 (p. 124.24-26): '... fieri uidetur aliquibus huius incisio secundum materiarum differentias. omnis enim particula corporis saluari uel iuuari nata est ex dieta uel farmacia uel cyrurgia.'

87 *PAM* 6.2 (pp. 38-39, 124-25). The Latin version omits aetiology.

88 *PAM* 6.3-6 (pp. 40-43, 125.3-28). According to the Empiricists the final parts of medicine are semiotics (subdivided into diagnostic and prognostic parts), hygiene (defined as the preservation of health), and therapeutics (which is subdivided into regimen, surgery, and pharmacology). The constitutive parts are αὐτοψία, ἱστορία, and μετάβασις τοῦ ὁμοίου, but Galen here, as elsewhere, points out that not all Empiricists agreed that the last is one of the constitutive parts of medicine. See n. 42 above and Deichgräber, *Empirikerschule* (n. 48 above) fr. 12-15, 17, 21, 62-63, 65 (pp. 90.24-91.11, 91.16-33, 93.16-20, 95.21-96.6, 97.7-11, 98.7-13, 100.18-20, 128.21-129.18, 129.35-130.12 and pp. 301-305.

89 *PAM* 7.1 (pp. 42-43, 126.4-5).

90 *PAM* 7.2 (pp. 42-43, 126.6-12). Both Ḥunain – or his son Isḥāq (n. 12 above) – and Niccolò seem to have had some difficulty rendering, and perhaps understanding, the well attested Galenic distinction between καθ'ἕξιν and κατὰ σχέσιν (the Arabic perhaps comes closer to the Greek by distinguishing between a 'firmly rooted' and a 'superimposed' healthy or diseased disposition; Niccolò used *secundum habitum* for καθ'ἕξιν and *secundum habitudinem* for κατὰ σχέσιν). For Galen's uses of this distinction see eg. *De sanitate tuenda* 5.4.2-3 (*CMG* V 4,2, pp. 142.23-143.1); *Thrasybulus* 7, 9, 31, 40 (*Scr. min.* III 40.2-23, 42.16-24, 74.20-23, 89.11-15); *De methodo medendi* 7.12, 8.3, 8.9, (X 518, 557, 598 K); *De compositione medicamentorum* 2.1 (XII 504 K). Cf. also the contrast between διάθεσις ἑκτική and διάθεσις κατὰ σχέσιν in *De locis affectis* 3.3 (VIII 140.14-16 K).

91 *PAM* 7.3 (pp. 42-43). On the Latin version see n. 92.

medicine – previously assigned to the neutral disposition – now in the twofold division become parts of the branch of medicine that deals with a less firmly rooted, more transient healthy disposition, ie. one that is more susceptible to disease but not diseased.[92] Galen did not, however, include paediatrics in this model; the paediatric part, which was correlated with the 'neutral' disposition in Galen's tripartite model, here seems to become sidelined by the return to a bipartite dispositional scheme. He offered no identification of the parts of medicine assigned to lasting or temporary diseased dispositions, adding only that they 'have no names, though some diseases are called chronic and others acute'.[93] More significantly, he here seems to have made no attempt to address the relative merits of the tripartite and bipartite dispositional models, let alone to address their partial incompatibility. Chapter 7.2-3 leaves the reader with the impression that Galen has come down firmly on the side of the bipartite model, yet 7.4 proceeds on the basis of the results initially achieved through the tripartite model, ie. the seven 'primary and greatest parts' of medicine, as though the transition between the two models is seamless.

The 'primary and greatest parts' of medicine are preceded by others 'without which these branches cannot exist'. Starting with the proximate antecedents of the 'greatest parts' of medicine, and moving to the ultimate antecedent parts, Galen introduced (1) the semiotic branch and its two subdivisions (expertise in diagnosis of the present and prognosis of the future),[94] and, 'on a par with' it,[95] (2) aetiology or the part concerning knowledge of causes of health and illness, including all the material causes, such as what comes into the body, what acts on it, what is evacuated from it, and what impinges on it externally.[96] The powers or properties (δυνάμεις) of all these have to be examined and known. Galen more than once insisted that both the semiotic and the aetiological branches precede the therapeutic and hygienic parts.[97] The semiotic branch as a whole in turn is 'resolved into [several] principles'

92 Lyons, *PAM* 7.3 (pp. 43.22-24): 'As for what deals with health in the second category [*scil.* with transient or κατὰ σχέσιν health], this comprises prophylaxis, the recuperative branch, and that of geriatrics'. Niccolò (p. 126.13-15) – or the Greek manuscript he used – seems to misrepresent the original. In his Latin version not only the 'euectic' but also the recuperative and geriatric parts are presented as subdivisions of the 'hygienic' branch, which is depicted as devoted to 'habitually healthy dispositions' and hence could not accommodate the recuperative branch. Moreover, in Niccolò's version no parts of the branch assigned to 'temporarily healthy dispositions' are identified, and the prophylactic part disappears entirely.

93 *PAM* 7.3 (pp. 42-43, 126.17-18). Like Aristotle, Galen was keenly aware of the problem of things without names (τὰ ἀνώνυμα); see eg. *De pulsuum differentiis* 1.5, 1.11 (VIII 508, 525 K); *De pulsuum causis* 2.7 (IX 81 K); *De symptomatum causis* 3.12 (VII 271 K); *De compositione medicamentorum per genera* 2.1 (XIII 460 K); *In Hippocratis De natura hominis comm.* 1.5 (*CMG* V 9, 1, p. 20.22-24); *De simplicium medicamentorum temperamentis ac facultatibus* 4.22, 6.4 (XI 699, 813 K); *De musculorum dissectione ad tirones* 12 (XVIIIB 951 K). Cf. *De methodo medendi* 2.2, 9.2, 10.11, 12.1 (X 81, 603-4, 730-31, 810-11 K).

94 *PAM* 7.4 (pp. 42-43, 126.19-24).

95 Lyons, *PAM* 7.5 (p. 45.8-10): 'But there is here another branch of medicine that is on a par with the diagnostic [semiotic; *significatiue* Niccolò, p. 127.6] and precedes the therapeutic. This is the knowledge of all the causes of health and illness...'.

96 *PAM* 7.5 (pp. 44-45, 127.6-11). Here too the terminology used in *PAM* is consistent with Galen's usage in other works. For his (not always uncritical) comments on the differentiation of τὰ προσφερόμενα from τὰ ποιούμενα, τὰ κενούμενα, and τὰ ἔξωθεν προσπίπτοντα, see eg. *Thrasybulus* 24 (*Scr. min.* III, pp. 62.14-63.7); *De sanitate tuenda* 1.15.5-10 (*CMG* V 4, 2, p. 36.7-30); cf. *De methodo medendi* 10.7 (X 695-96 K). At times he implies that this fourfold distinction was invented by precursors. See n. 128 below.

97 Eg. *PAM* 7.4, 7.5 (pp. 42-45, 126.19-20, 127.1-5).

or starting points (ἀρχαί? *principia*, Niccolò) of its own,[98] which are identified as yet further antecedent parts. First, at the next remove lie three parts of medicine that precede the semiotic branch: (3) knowledge of all the parts of the living being (discovered by dissection of the dead), ie. anatomy; (4) the part of medicine concerning the activities or actions (ἐνέργειαι?) of bodily parts; and (5) the part that studies their uses or functions (χρεῖαι?). Preceding all of these, in turn, are: (6) the part concerning knowledge of the elements of the body (τὰ τοῦ σώματος στοιχεῖα); (7) the part concerning the blendings (temperaments, αἱ κράσεις); and (8) the part concerning the natural and psychic faculties (discovered through the vivisection of animals).[99] Parts (3)-(8) therefore are all presented as necessary but not sufficient conditions of semiotics, which in turn, along with aetiology, is a necessary but not sufficient condition both of the preservation of healthy dispositions (hygiene and the euectic, prophylactic, etc., parts) and of the therapeutic treatment of temporarily and chronically diseased dispositions (therapeutics and all its branches).

After multiple refinements and elaborations of his theory of the proper division of the medical *techne* Galen appears to have arrived at his final view: 'This, then, is the way to grasp the *techne* of medicine in accordance with its own special material.'[100] Yet, as so often when Galen's definitive view finally seems to come into sight after a series of detailed qualifications and refinements, here too further complicating perspectives instantly occurred to him. A distinction between the 'primary' and 'secondary' material or subject matter of medicine – or between the material most peculiarly its own and material that belongs to it only accidentally – now is brought into play. Health and disease, he added, are the primary subject matter of medicine, ie. the matter most peculiarly its own, but inasmuch as health and disease are dispositions of the body, the body itself becomes the secondary or accidental material of medicine.[101]

A consequence of the body being material for the medical *techne* is, according to Galen, that this *techne* 'can be split into as many parts as those into which the body can be divided'.[102] Having just reminded the reader of the division based on dispositions, Galen here explicitly reverted to the distinction between elementary and instrumental parts, thus re-animating the analogy between dissection and the division of medicine into its parts. Galen's hospitable

98 *PAM* 7.4 (pp. 44-45, 127.1-2).

99 *PAM* 7.4 (pp. 42-45, 126.24-127.7). For 'blendings' or 'temperaments' ('compounding' Lyons p. 45.1) Niccolò (p. 126.30) reads *crisibus*. Kollesch, Nickel, and Strohmaier (app. crit. to *PAM* p. 126.30) plausibly suggest that *crisibus* is a corruption of *crasibus* (or that the corruption of *kras.* to *kris.* already appeared in Niccolò's Greek manuscript).

100 *PAM* 8.1 (pp. 44-45 Lyons). Niccolò's translation (p. 127.12-13) conveys the same general sense as the Arabic but suggests a different wording in the original: *que quidem igitur <secundum* (add. B) > *propriissimam materiam medicatiue inuentio partium eius, ex tali*. See n. 107 below.

101 *PAM* 8.1-2 (pp. 44-45, 127.14-16). Schöne, *Galenus* (n.8 above) 35 n. on l. 300 plausibly reconstructs the Greek as follows: τὸ τοῦ ἀνθρώπου σῶμα τῆς ἰατρικῆς τέχνης ἐστὶν ὕλη δευτέρα καὶ ὡς ἂν εἴποι τις κατὰ συμβεβηκὸς τῷ τὰς τούτου διαθέσεις εἶναι ὑγίειαν καὶ νόσον. According to Galen, the body belongs to the secondary material of medicine by virtue of its susceptibility to disease and to being rendered healthy again by treatment, whereas, in the case of natural philosophy, it is by virtue of being generated and corruptible that the body belongs to its secondary material, ie. inasmuch as the primary material of natural philosophy includes generation and corruption. For Galen's use of ἡ δευτέρα ὕλη in a similar but not identical sense see *De crisibus* 2.9 (IX 678 K = 150.1-5 Alexanderson).

102 Lyons, *PAM* 8.3 (p. 45.33-34).

response to several divergent principles of division by now suggests a model according to which the number of parts of medicine will correspond to the number of 'element-like' and 'instrumental' parts of the body, to the number of diseases occurring in each of these parts (for example, a cataract in the 'instrumental' part called eye), and to the 'elemental' parts into which each *techne* of treating a disease in each of these parts can be divided. Moreover, if a disease has to be treated with drugs, the pharmacological part of the therapeutic branch of the *techne* will similarly have to be divided into its elementary parts, such as the sciences of the simples and the sciences of applicable compound drugs. Similarly, the dietetic part of the therapeutic *techne* has to be divided into the parts of regimen that are pertinent to conditions of various elementary or instrumental parts of the body,[103] etc.

Galen was quick to remind his audience, however, that the body, as the secondary material of the medical *techne*, is not the only or primary starting point for division. Dispositions, simple and compound diseases, general therapeutic principles (for example, 'opposites cure opposites'), natural and unnatural bodily conditions, etc. all can also serve as legitimate starting points for the division of medicine.[104] Accordingly, said Galen:

> The lesson of this is that, wherever you start your division, if you follow it properly until you reach the ultimate and element-like sciences, you will find that it will bring you to those we have already mentioned.[105]

The purpose of these many different divisions is to bring out 'all the subject comprises without any omissions.'[106] One therefore should not be perturbed by the multitude of divisions, Galen concluded, nor by the variety of methods of division. Instead one should apply the criterion of completeness to any account of the parts. And, Galen reiterated in a final reaffirmation of his view of correct division, in one's own attempts at dividing medicine into its parts:

> the best course is ... that in your division you should begin from whatever may be the most proper material and then you should add in all the other classes of division there may be, using the method that I have mentioned.[107]

103 *PAM* 8.3 (pp. 44-47, 127-28).

104 *PAM* 8.4-9.1 (pp. 46-49, 128). The bodily conditions that are contrary to nature are in turn subdivided into those that are diseases, those that are symptoms of diseases, and those that are causes of diseases (9.1). The tripartition is similar to, but not co-extensive with, the distinction between body, cause, and sign in *Ars medica*; see above, n. 72.

105 Lyons, *PAM* 8.7 (p. 47.29-32). Niccolò's translation reads as follows (p. 128.18-21): 'quo et manifestum, quoniam, undecunque inceperit quis incidere, si recte incidit usque ad ultima et elementalia theoremata, peruenit ad scientiam eorum, que dicta sunt.'

106 Lyons, *PAM* 9.6 (p. 49.32-33); cf. Niccolò (p. 129.14-15): '... sed si nihil eorum, que occurrunt, obmissum est, considera'.

107 Lyons, *PAM* 9.6 (p. 49.34-37). Niccolò's translation (p.129.16-18) here corresponds closely to the Arabic version: 'optimum est autem ... principium facere a propriissima materia, et deinde alia secundum prius dictam methodum adicere'. On 'most proper material' or 'its own very special material' (Lyons; Niccolò: *propriissima materia*; ὕλη ἰδιωτάτη?) see also *PAM* 8.1 (n. 100 above). Galen is not averse to using the superlative of ἴδιος, eg. the hand is the ὄργανον ἰδιώτατον of a human being (*UP* 17.1; II 442.2 Helmreich) and extreme thirst is τὸ σύμπτωμα ἰδιώτατον of the 'burning fevers' (καῦσοι πυρετοί: *In Hp. Epid. III. comm.* 3.35; *CMG* V 10,2,1, p. 133.15 Wenkebach). But to my knowledge ἰδιωτάτη ὕλη does not appear in Galen's works that are extant in Greek. See n. 100 above.

Faced with the enormous complexity and variability of the human body, with the diversity of diseases, with many temporary or chronic deviations from the healthy norm, with the great variety of resources in the good physician's therapeutic arsenal, and with the numerous different interventions physicians could undertake at different times in response to different bodily disorders (or to the same disorder in different bodies), Galen seems to have concluded that numerous different kinds of professional expertise constitute the medical *techne*. In part for this reason he not only offered a critical review of prior divisions but proposed many of his own, heaping division upon division. Doing so, he believed, is necessary for identifying the multiple, divergent components of the medical *techne* and for understanding the relations between them. In short, as long as the division or 'cutting' was, in his view, done correctly, Galen cheerfully proliferated the divisions of the *techne* and hence the number of its parts. And the correct method, as indicated above, always begins with the ὕλη ἰδιωτάτη of the thing being divided, and always ends in the discovery of the ultimate and most elemental forms of scientific knowledge (*elementarissimae scientiae, ultima et elementalia theoremata*, Niccolò[108]) concerning this most proper material – i.e. of the material most distinctively belonging to the thing that is being divided. Only by reaching knowledge of the fullness of its parts, in other words, will one arrive at the science of medicine. But should there also be distinct specialists devoted exclusively to each of these numerous distinct branches of the medical *techne*?

5. Specialization and urbanization

Already in the second chapter Galen acknowledged that the discovery of the correct division of medicine into its parts not only is a formidable theoretical challenge but also has practical implications, for example, for the question whether and to what extent specialists practising a single branch of medicine are medically desirable and economically viable.[109] Before I turn to Galen's response to this question, a brief allusion to some historical contexts may be useful.

Specialized medical expertise had a long pre-Galenic history.[110] The evidence concerning

108 *PAM* 9.4 (p. 129.8): '... necesse est incisionem desinere usque ad elementarissimas scientias'; 8.7 (p. 128.19-21): '... si recte incidit usque ad ultima et elementalia theoremata, peruenit ad scientam eorum, que dicta sunt'; 8.3 (p. 127.30-31): '... harum enim scientie sunt elementales particule totius artis'; 8.4 (p. 128.2): '... ex huius materie incisione usque ad ultimas et elementales artis particulas accedit'. Cf. Niccolò's use of *paruissime particule artis* in 3.1 (p. 121.5) and of *(artis) minime particule, ultime et minime et simplices noticie*, and *minime et elementales particule*, eg. in 3.1-2, 4.2, 8.6 (p. 121.7-11, 122.17-18, 128.10-11). Given the literal nature of Niccolò's translation, these variations probably reflect different emphases in Galen's descriptions of the στοιχεῖα αἰσθητά, eg. τὰ στοιχειώδη θεωρήματα, τὰ στοιχειώδη μόρια τῆς τέχνης, τὰ ἐλάχιστα μόρια, τὰ ἐλάχιστα καὶ ἁπλᾶ/ἁπλούστατα μόρια τῆς τέχνης. Niccolò's *elementarissimas* (p. 129.8) might reflect the superlative στοιχειωδέστατος, which is attested for Galen, but in a different context: *De praesagitione ex pulsibus* 2.1 (IX 271 K), περὶ ... τῶν πρώτων τε καὶ ὡς ἂν εἴποι τις στοιχειωδεστάτων σφυγμῶν ... See also nn. 52, 57, 58, and 59 above.

109 By 'specialist' I mean a physician whose practice is limited to some special, distinctive branch of medicine, and by 'specialization' I mean a medical practice limited to one such branch.

110 Although Egyptologists such as Wolfhart Westendorf in recent years have cautioned that early Egyptian medical papyri do not offer unequivocal evidence of medical specialists practising only one branch of medicine, there is evidence at least of a very early recognition in Egypt of different types of medical expertise, even if a single physician could become famous for several forms of expertise (eg. a court physician of the Old Kingdom, Iri, is identified on his tombstone as 'eye-doctor of the palace, belly-doctor of the palace, shepherd of the anus', and, more generally, as 'doctor of the palace'); see Westendorf, *Erwachen der Heilkunst. Die Medizin im alten Ägypten* (Zurich 1992) 237-38. See also P. Ghalioungui, 'Early specialization in ancient Egyptian medicine and its possible relation to an archetypal

medical specialization in the archaic and classical periods of Greece admittedly is murky (with the exception of allusions in the classical period to figures such as midwives and sellers of medicines, who tend, however, not to be identified as *iatroi* [111]). But the Hellenistic and Roman periods offer less ambiguous evidence of the development of separate, distinct groups of physician-specialists, even as the unity of the entire *techne* kept on being proclaimed by some medical writers.[112] The rise of specialization was accompanied by an increasingly animated debate about the proper division of medicine into its parts,[113] and it is to this historical context that some of Galen's remarks on specialization belong.

Surgery, in particular, seems to have become a field of specialization no later than the Hellenistic period. Markwart Michler has plausibly argued that this development went beyond the appearance of a number of surgeons to include the rise of several surgical subspecialities in the second and first centuries BC.[114] Several such Hellenistic sub-specialities were still known to Galen as having achieved professional autonomy, for

image of the human organism', *MedHist* 13 (1969) 383-86, who emphasizes that specialists are a minority of the physicians mentioned in medical papyri of the Pharaonic period. At the time of Persian rule in Egypt Herodotus (2.84) reported that in Egypt 'each healer is a healer for a single disease ...; some are established as healers for the eyes, some for the head, some for the teeth, some for the belly, and some for invisible diseases'. He implied that this widespread, largely anatomically defined specialization – which bears a striking resemblance to the situation Galen describes in second-century Rome (see below) – is a distinctively Egyptian, un-Greek phenomenon.

111 Ancient scholiasts already remarked on the differences in the medical skills ascribed to the Homeric warrior-healers Podalirius and Machaon; see eg. Sch. Il. 11.515; Eust. *Comm. Il.* 859.38-44 (vol. 3,243-44). On midwives in the classical epoch see eg. S. *Alexandros* F 99.1 (*TrGF*, IV p. 148 Radt); Pl., *Tht.* 149a-151c, *Plt.* 268b; Arist. *HA* 9(7). 10.587a 9; Lys. (?) *Contra Antigenem de abortu,* fr. 333.1 Thalheim (ed. maior, 2nd ed. 1913). On the ἀκεστρίδες, 'healing women' who assisted women in labour see Hp. *Carn.* 19.6 (VIII 614.11 Li. = p. 22.16 Deichgräber = 202.26 Joly); on the female 'cutter of the umbilical cord' (ἡ ὀμφαλητόμος) see Hp. *Mul.* 1.46 (VIII 106.7 Li.) and, as early as the sixth century BC, Hippon. 33 Degani (= 19 West = fr. 12 Diehl). The later Greek lexicographic tradition (Hesychius, *Suda*, Photius) consistently identified ἡ ὀμφαλητόμος with 'midwife' (μαῖα). In a funerary epigram of the second half of the fourth century BC the Acharnian Phanostrate is commemorated as μαῖα καὶ ἰατρός (*IG* II/III^2 6873). For references from the classical period to sellers of medicines see eg. Ar. *Nu.* 767, *Pl.* 884; Critias 88B70 D/K; Aeschin. 3.162. The much vaunted medical root-cutters (ῥιζοτόμοι) or herbalists, who also are not identified as ἰατροί, probably built on a long tradition, but not until the time of Theophrastus are they explicitly identified as a distinct group of specialists. See also D. Nickel, 'Berufsvorstellungen über weibliche Medizinalpersonen in der Antike', *Klio* 61 (1979) 515-18; N. Demand, 'Monuments, midwives and gynecology', in *Ancient medicine in its cultural context*, ed. Ph. J. van der Eijk, H. F. J. Horstmanshoff, P. H. Schrijvers, 2 vols. (Amsterdam/Atlanta 1995) 275-90; J. Korpela, '*Aromatarii, pharmacopolae, thurarii et ceteri.* Zur Sozialgeschichte Roms', *ibid.* 101-18; H. King, *Hippocrates' Woman. Reading the female body in ancient Greece* (London/New York 1998) 177-80.

112 On medical specialists in the Greco-Roman world see H. Gummerus, *Der Ärztestand im römischen Reiche nach den Inschriften* (Helsingfors 1932); G. Baader, 'Spezialärzte in der Spätantike', *MHJ* 2 (1967) 231-38; M. Michler, *Die hellenistische Chirurgie, Teil I: Die alexandrinischen Chirurgen* (Wiesbaden 1968) 138-55; *id., Das Spezialisierungsproblem und die antike Chirurgie* (Bern 1969); P. Platiel, *Das Spezialistentum in der Medizin bis zum Ausgang der Antike* (Munich 1977); K.-D. Fischer, 'Der ärztliche Stand im römischen Kaiserreich', *MHJ* 14 (1979) 163-66; P. Mudry, 'Médecins et spécialistes. Le problème de l'unité de la médecine à Rome au Ier siècle ap. J.-C.', *Gesnerus* 42 (1985) 329-36; J. Korpela, *Das Medizinpersonal im antiken Rom: eine sozialgeschichtliche Untersuchung*, Annales Academiae Scientiarum Fennicae: Dissertationes Humanarum Litterarum 45 (Helsinki 1987) 30-32, 49-51, 81-85, 97-99, 123-25, 135-36, 143-45.

113 See n. 45 above.

114 Michler, *Spezialisierungsproblem* (n. 112 above) 28-31.

example, cataract surgery, bladder stone surgery, wound surgery, and orthopaedic surgery.[115] Given the fragmentary nature of the Hellenistic evidence, it remains less than clear, however, whether any early Hellenistic physician actually restricted his or her practice to a single form of specialized expertise and, if so, how many such specialists and specialities there were. Yet there is little doubt that some physicians became known only for the practice of a single, relatively narrow expertise, such as those described by Galen in *PAM*.

In the century before Galen the development of surgery into a relatively autonomous area of professional expertise is confirmed by Aulus Cornelius Celsus. In the preface to his two books on surgery, which draw heavily on Hellenistic sources, he described Hippocrates as a great surgeon but the rise of surgical specialists as a post-Hippocratic development:

> Although it is very ancient, this part [*scil.* surgery] was cultivated more by the father of all medicine [*scil.* Hippocrates] than by his predecessors. Subsequently it was separated from the other parts and began to have its own professional experts.[116]

Not unlike Galen, Celsus nevertheless struggled to define the boundary between surgeons and other practitioners:

> It could be asked, however, what should be claimed by this part [surgery] as its proper province, because surgeons (*chirurgi*) claim for themselves treatments of wounds as well and of many ulcerations – treatments which I have described elsewhere [*scil.* in the non-surgical books of *Medicina*].[117]

Furthermore, like Galen, Celsus resisted specialization:

> I, for one, state expressly that the same person can indeed accomplish all these things; and when divisions [*scil.* of medicine into its parts] are made, I praise him who has grasped as much as possible.[118]

The Roman encyclopaedist in fact insisted that surgery cannot do without the other two major branches of therapeutics, ie. without regimen and pharmacology.[119]

In *PAM* Galen explicitly recognized that oculists may have a special historical claim to practising a legitimate medical speciality, and, as Schöne suggested, *PAM* may even be addressed to an eye-doctor.[120] Galen's acknowledgement of the long history and special status of eye specialists reflects an unusual, perhaps unique historical development. Its Hellenistic

115 Some Hellenistic orthopaedic surgeons became especially famous for using an elaborate new mechanical technology that drew on Hellenistic technical innovations in siege artillery; see von Staden, 'Andréas de Caryste et Philon de Byzance: médecine et mécanique à Alexandrie', in: *Sciences exactes et sciences appliquées à Alexandrie* (*III siècle av. J.-C.--I siècle ap. J.-C.*), Actes du Colloque international de Saint-Étienne, publiés par Gilbert Argoud et Jean-Yves Guillaumin (Centre Jean-Palerne, Mémoires XVI: Université de Saint-Étienne, 1998) 147-72.

116 Cels. 7 pr. 2-3 (*CML* I, p. 301 Marx).

117 Cels. 7 pr. 5 (p. 302 Marx). Andreas (n. 115 above), a Ptolemaic court physician, is an example of the Hellenistic orthopaedic surgeons who did not confine their practice to this specialty. Among later physicians who are described both as ἰατροί and as χειρουργοί was T. Aelius Asclepiades, a second-century surgeon of the *Ludus Matutinus*; see *IG* XIV 1330; *IGRR* I 182; *IGUR* 282; Gummerus, *Ärztestand* (n. 112 above) 42 no. 138; Korpela, *Medizinpersonal* (n. 112 above) 191 no. 206.

118 Cels. 7 pr. 5 (p. 302 Marx).

119 Cels. 7 pr. 1 (p. 301 Marx).

120 See n. 49 above.

roots remain obscure, but starting in the early Roman empire the popularity of the *ocularii* or *ocularii medici* (οἱ ὀφθαλμικοί), many of whom seem to have been itinerant dispensers of eye salves and of other drugs for a great variety of eye ailments, is attested not only by literary sources but also by numerous inscriptions and by the discovery of countless oculists' collyrium-stamps, ie. small incised blocks of stone used to mark sticks of eye salve with the names of their owners or inventors, with an identification of the eye salves, and occasionally with the name of the ailment for which a given salve was used.[121]

The sizable caches of cataract needles discovered in Roman Gaul and in other parts of the Greco-Roman world[122] have been interpreted as evidence that at least some of the itinerant oculists also performed cataract operations. In this context it is striking that Galen, as indicated above, used couching cataracts as a paradigm case of a distinctive, autonomous medical specialty that itself can be further divided into various subspecialities. Medical instruments, including cataract needles, found with collyria-stamps in the tombs of *medici ocularii* suggest, however, that the specialist who couched cataracts also treated a variety of other ailments.[123] Be this as it may, no later than the first century BC surgery and eye therapy were recognized as distinct forms of expertise practised by distinct classes of specialists, even as some physicians claimed for themselves not only one of these two forms of expertise but also much broader medical competence.[124]

Surgery and ophthalmology were not the only branches of medicine that drew specialist-practitioners in the early Roman empire. Soranus of Ephesus, for example, refers to physicians called γυναικεῖοι because they specialized in treating the affections (πάθη) of women, and he distinguishes these gynaecologists from midwives.[125] Voices that deplored the rise of medical specialization were raised already in the first century BC, however, also

121 See E. Espérandieu (ed.), *CIL* XIII, pt. 3, fasc. 1 (10021.1-10229); *id.*, 'Nouveaux cachets d'oculistes', *RA* ser. 5 vol. 26 (1927.2) 158-169; G. Ovio, *L'oculista ai tempi dell' impero romano* (Rome 1940); V. Nutton, 'Roman oculists', *Epigraphica* 34 (1972) 16-29; G.C. Boon, 'Potters, Oculists, and Eye-Troubles', *Britannia* 14 (1983) 1-12; R. Boyer et al., 'Découverte de la tombe d'un oculiste à Lyon (fin du IIe siècle après J.-C.). Instruments et coffret avec collyres', *Gallia* 47 (1990) 215-49; M.H. Marganne, *L 'ophthalmologie dans l'Égypte gréco-romaine*. Studies in Ancient Medicine 8 (Leiden/New York/Cologne 1994); R. P. Jackson, 'Eye medicine in the Roman Empire', *ANRW* II 37.3 (Berlin/New York 1996) 2226-2251.

122 See eg. M. Feugère, E. Künzl, U. Weisser, 'Les aiguilles à cataracte de Montbellet (Saône-et-Loire). Contribution à l'étude de l'ophthalmologie antique et islamique', *JRGZ* 32 (1985) 436-508, republished (in a separate volume) by the *Societé des Amis des Arts et des Sciences de Tournus* 87 (1988); *id.*, 'L'opération de la cataracte dans le monde romain et l'instrumentarium de Montbellet', *DossArch* 1988 no. 123 67-71.

123 See Boyer, 'Découverte' (n. 121 above) 224-243; Nutton 'Roman oculists' (n.121 above) 23-29.

124 Herophilus and Demosthenes Philalethes, for example, were famous as eye-doctors but also claimed expertise in other branches of medicine; see von Staden, *Herophilus* (n. 45 above) 160-61, 202-06, 237-39, 252-55, 317-18 (fr. 140a), 423-26, 570-74, 582-84. Galen and Alexander of Tralles are later examples of physicians well known for their ophthalmological expertise, yet far from limited to this speciality. See also Nutton, 'Roman oculists' (n. 121 above) 25 on the oculists as itinerant doctors not restricted to treating eye ailments; Michler, *Spezialisierungsproblem* (n. 112 above) *passim*.

125 Sor. *Gyn.* 3.3.1 (*CMG* IV, p. 95.7-8 Ilberg) = 3.1.32-33 (III p. 3 Burguière/Gourevitch/Malinas). Soranus affirms, however, that even midwives must be trained in all parts of medicine, given that some disorders of women 'must be treated by regimen, some by surgery, and some must be straightened out by drugs': *Gyn.* 1.4.2 (*CMG* IV p. 5.10-13 Ilberg) = 1.1.25-30, 1.3.10-11 (I pp. 5, 7 Burguière/Gourevitch/Malinas). See n. 47 above on this threefold division.

in non-medical literature.[126] Galen shared this sentiment. He nevertheless insisted on the identification of all branches of medicine, down to its smallest elements, and in this context he repeatedly referred to specialist practitioners. Two of the more extensive surviving ancient enumerations of specialists occur in his *Thrasybulus* and *On the Parts of the Medical Techne*. According to the latter, all kinds of specialists could be found practising medicine in Rome, including those called dental doctors (ἰατροὶ ὀδοντικοί), ear doctors (ὠτικοί), anus doctors (ἑδρικοί?), eye doctors (ὀφθαλμικοί), couchers of cataracts (παρακεντηταί), hernia cutters (κηλοτόμοι), and stone removers (λιθοτόμοι).[127] In his *Thrasybulus* Galen mentioned almost all of these as well as specialists in regimen (διαιτητικοί), specialists in treatment by means of drugs or plants (φαρμακευτικοὶ ἢ νὴ Δία βοτανικοί), doctors called 'wine-givers' (οἰνοδόται) or 'givers of hellebore' (ἐλλεβοροδόται).[128]

Galen conceded that, given the well-established tradition of specialists in treating eye ailments, proponents of a sharper division of expert labour might reasonably argue that there should likewise be specialists for treating teeth or ears and every other part of the body, so that there will be as many types of specialists as there are parts of the body.[129] Moreover, he argued that since specialists in treating specific disorders (cataracts, hernias, stones) are also recognized – experts of each such kind having their own, distinct, professional designation (coucher, hernia-cutter, stone-cutter) – there will in fact be even more physicians than there are parts of the body, at least if this is upheld as a paradigm for a professional division of

126 Eg. Cic. *de Orat.* 3.32 depicts Crassus as saying: 'Not in this matter alone but also in several others the greatness of the arts has been diminished by the division and separation of their parts. Or do you really think that at the time of the great Hippocrates of Cos there were some physicians just for treating diseases, others for treating wounds, and still others for treating the eyes?' The argument against medical specialization here is extended not only to the unity of mathematics, of music, and of literature but to a universal embrace of all branches of culture.

127 *PAM* 2.2 (pp. 26-29, 121.17-27). Galen's use of the corresponding Greek terms is independently attested; see eg. *Thrasybulus* 24 (*Scr. min.* III 62.3-4): ... τούτους ὀφθαλμικούς τε καὶ ὠτικοὺς καὶ ὀδοντικοὺς ἰατροὺς ὀνομάζουσιν. On τῶν ὀφθαλμικοὺς ἑαυτοὺς ὀνομαζόντων ἰατρῶν see eg. Gal., *UP* 10.11 (II 91.18-20 Helmeich) and *De differentiis febrium* 2.16 (VII 392 K). On eye-specialists (ὀφθαλμικοί) also see Gal., *Thrasybulus* 24 (*Scr. min.* III 64.10-21), *De methodo medendi* (X 941, 1019 K), and *In Hippocratis Aphorismos comm.* 6.31 (XVIIIA 47-49 K). In *De compositione medicamentorum secundum locos* 3.1, 4.7, 4.8 (XII 632, 732, 745, 746, 751, 754, 760, 763, 766, 771, 786 K) Galen refers by name to several ὀφθαλμικοί, eg. Zoilus, Capito, Hero, Sergius the Babylonian, Hermeias, Paccius, Gallio, Gaius, and Axius. It is noteworthy that most of these ὀφθαλμικοί have names that are not Greek in origin, whereas most ἰατροί have Greek names. The epigraphic evidence reflects the same tendency: there are numerous Latin inscriptions that refer to oculists but few in Greek. In fact, the only Greek example I can recall is bilingual: the Latin and Greek epitaph of a first-century BC ἰατρὸς ὀφθαλμικός /*medicus ocularius*, Gaius Terentius Demosthenes, on a marble stele found in Caesarea in Mauretania; see *CIL* VIII 21105; *IGRRI* I 299; Gummerus, *Ärztestand* (n. 112 above) 82 no. 313. For the specialists known as παρακεντητής, κηλοτόμος, and λιθοτόμος see also *Thrasybulus* 24 (*Scr. min.* III 61.25-62.2).

128 Gal., *Thrasybulus*. 24 (*Scr. min.* III 61-65). As in *PAM*, Galen here concedes that further divisions might be justifiable, for example one could divide expertise in dietetics into knowledge of things applied to the body, things evacuated from it, things done or effected, and things impinging upon it from outside (τὰ προσφερόμενα, κενούμενα, ποιούμενα, ἔξωθεν προσπίπτοντα); see n. 96 above. Each of these can also reasonably be further subdivided, for example, the *techne* concerning τὰ προσφερόμενα into the *technai* of applying (a) this or that drug or (b) food or (c) drink. While Galen grants that each of these subdivisions represents a distinct *techne*, he also insists on their unity: they are all parts (μόρια) of a larger *techne*, united by a shared goal (σκοπός), namely, health.

129 *PAM* 2.2 (pp. 26-29, 120).

labour: 'For not only will there be a single specialist to deal with each part of the body individually, but each of the ailments of each of the parts will have its separate specialist.'[130]

In view of the enormous complexity of the human body, the diversity of diseases, and the vast scope of medical knowledge, it is, in Galen's view, not surprising that medicine in Rome has been divided into so many branches that a specialist can be found for any ailment of any bodily part that requires medical attention.[131] It is striking that Galen in this context presented the only constraint on a proliferation of specialists as socio-economic. Specialization, he argued, was not economically viable in small cities or rural areas, whereas it could be sustained in large urban centres:

> For, in a small town someone who [only] couches eyes or cuts hernias will not be sustained at all, whereas in Rome and Alexandria, thanks to the great multitude of inhabitants, there are enough people to ensure a livelihood for those who practice any single [specialized] branch of medicine whatsoever, let alone for those who have more medical competence than that.[132]

This does not mean that there were no specialists outside the largest urban centres; rather, as Galen confirms, 'elsewhere, ... where towns are small, the single branch specialists must travel continuously from place to place' in order to make a living.[133] The itinerant specialist, well attested in the case of the ancient oculists, faced a formidable challenge, Galen claimed: he would have to cover the territory of all of Greece to find as many patients as a specialist could find in a single large city such as Rome.[134]

According to Galen, then, the multiplicity of specialists, each practising a small part of the *techne*, is within certain limits a viable practical consequence of massive urbanization, even if the best physician will have knowledge of all branches of medicine, including its smallest, 'elemental' parts (not to mention all of philosophy). Highly specialized *practice* is depicted neither as flowing from theoretical considerations nor as practically desirable, but rather as a socio-economic fact of urban life: it is conditioned by market factors that determine economically sustainable levels of the division of expert medical practice. By contrast, the multiplicity of *theoretical* divisions of medicine is presented as a scientific and methodological necessity. Dividing medicine into its branches, all the way to its smallest in-divisible elements or parts, is a necessary though not sufficient condition for gaining the medical *episteme* on which an efficacious practice of one's *techne* must be based, whether one practises in Rome or in a small town. Galen's valorization of a scientific knowledge of all the parts of the medical *techne*, in short, is presented as an epistemological and pragmatic necessity that is, in principle, independent of demographic and socio-economic considerations.[135]

Institute for Advanced Study, Princeton

130 Lyons, *PAM* 2.2 (p. 29.10-12). Niccolò's rendering (p. 120.28-29) is more compressed than the Arabic: '... plures erunt quam corporis particule medici; secundum enim unamquamque passionem erit unus medicus'.

131 *PAM* 2.3 (pp. 28-29, 120.29-31).

132 *PAM* 2.3 (pp. 28-29, 120).

133 Lyons, *PAM* 2.3 (p. 29.21-23); see pp. 120.34-121.1 for Niccolò's version (which presents a crux).

134 *PAM* 2.3 (pp. 28-29, 120.32-121.3).

135 I am grateful to J. Chr. Bürgel for his expert help with points of significant detail in the Arabic version of *PAM*, and to Vivian Nutton for valuable suggestions.

THE CONTRIBUTION OF GALEN, *DE SUBTILIANTE DIAETA (ON THE THINNING DIET)*[1]

JOHN WILKINS

De subtiliante diaeta (On the Thinning Diet) is a short treatise which appears to draw on a number of characteristic concerns of Galen in his work on nutrition. It is practical and tied to the normal diet in the ancient Mediterranean, as we understand it;[2] it is directed mainly at the professional doctor; it derives from a considerable body of earlier theory; and it concerns itself with problems of terminology and definition. Galen allows himself some short digressions but for the most part is well focussed on the topic. This last point is worth stressing since many treatises on diet in antiquity, for example Galen's own *De alimentorum facultatibus,* aim for completeness and list a wide range of qualities in each category of food, often making it difficult to distinguish particular physiological effects. This treatise, by contrast, is concerned only with the thinning of humours and is abundantly clear.

The treatise was not available to Kühn in a Greek text, which came to the West only in 1840.[3] But it had been cited by Oribasius in the fourth century; translated into Arabic in the ninth century, and apparently used by Maimonides in the twelfth; and translated into Latin in the early fourteenth century by Niccolò da Reggio.[4]

De subtiliante diaeta appears to be a comparatively early composition. Galen mentions it in eight other works, including *De sanitate tuenda, De alimentorum facultatibus* and *De bonis malisque sucis.* That is not to say, of course, that Galen's views on the subject were necessarily destined to change or mature. They do not, at least, appear to develop greatly between *De subtiliante diaeta* and *De alimentorum facultatibus.*

I noted above the long history of dietetics on which Galen draws: one would not be aware of them from the treatise itself. Notably for a treatise by Galen, the work is free of polemic. I am not sure how this is to be interpreted. The three possibilities are these:

1. There had been no systematic treatment of the thinning diet before Galen, and hence he had no one to argue against. This seems unlikely. Galen refers at the beginning (433.16-7)

[1] Also known as *De victu attenuante.* It is cited here by the pages of the edition by Karl Kalbfleisch, *CMG* V 4,2 (Berlin 1923) 432-51. An English translation is given by P. N. Singer, *Galen. Selected Works* (Oxford 1997) 305-24.

[2] On the ancient diet and its differentiation across social classes see P. Garnsey, *Food and Society in Classical Antiquity* (Cambridge 1999).

[3] On Parisinus suppl. grec 634 see Kalbfleisch, *CMG* V 4,2, xlviii-lvii; V. Boudon, *Exhortation à la Médecine, Art médical* (Paris 2000) 211-14. The *editio princeps* was made by K. Kalbfleisch (Leipzig 1898).

[4] On the Arabic version see Kalbfleisch (1923), on the Latin version of Niccolò da Reggio, see N. Marinone, ed., *Galeno: la dieta dimagrante* (Turin 1973); V. Nutton, ed., *Galen, On prognosis, CMG* V 8,2 (Berlin 1979) 23-26.

to the 'thinning' and 'thickening' diets as recognised medical terms. Even if there were no medical texts on the topic, the phenomenon was sufficiently well known among doctors for Galen to develop counter-arguments on certain points if he so desired.

2. There were predecessors, but Galen chose not to engage with them.

3. Predecessors had written about the thinning diet within more wide-ranging treatments on diet. Galen thus waited to take them on in the larger context of *De alimentorum facultatibus.* We should note, however, that the level of polemic in this longer work is not extensive and is largely confined to mistaken classifications of plants by Mnesitheus of Athens and to the errors of the Atticists in obscuring technical terms.

It seems sensible to conclude that the treatise was either designed for use as a practical handbook, for which polemic would be a distraction, or that Galen saw no gross errors that were in need of correction on this topic. There appear to be errors that he might have wished to correct, such as the views of Diphilus of Siphnos on the turnip.[5] Diphilus claims that 'the turnip (*gongulis)* is thinning, acrid and difficult to digest. It also causes much wind.' Galen's view, conversely, is that the turnip is very unsuitable for the thinning diet (444.14-21).[6]

The treatise is apparently ordered according to two principles. The first promotes the best foods for the thinning diet (the alliums) to the beginning and the worst (milk and cheese) to the end. This order resembles that of *De alimentorum facultatibus* in which each category (cereals, other plants, animals) is ordered according to foods most frequently consumed. The second principle appears to follow the order of food as it was eaten in a meal. I return to this order below.

The listing and classification of foods appear to be Galen's main purpose. No general *argument* is made about the thinning of humours, or about which humours he has in mind. It is, however, made very clear that thick, sticky and cold humours cause the disorders that are under consideration: phlegm appears to be the major problem, and phlegm-producing foods are frequently noted. Blood is also sometimes considered.[7] The texture of the food is significant for its effect in digestion and for its suitability in the thinning action. Goat's meat and beef, for example, are said to have no harmful quality but are hard and comparatively difficult to assimilate, whereas fish are soft and friable.[8] Cooking also has a part to play in modifying texture.

[5] Quoted by Athenaeus, *Deipn.* 9.369d.

[6] Compare *De alimentorum facultatibus* 323.19-25 Helmreich, where we are told that the turnip distributes a comparatively thickening juice into the body.

[7] Compare *De alimentorum facultatibus* 364.17-365.2 Helmreich: 'nourishment (*trophe*) from the sea-bass and from other fish is generative of blood, and a blood that is lighter (*leptoteron*) in its formation than that formed from land animals. We use the term *leptoteron* sometimes for a comparison between two things, and sometimes, simply, without a comparison, for the average kind of blood that is between extremes. The extreme of badness in its formation is thick blood, like wet pitch; then there is whey-like blood that when it flows from the vein and thickens has much damp and watery content sitting on the surface. The best is precisely between these two, which comes from the finest made bread, which I wrote about in book 1, and from the winged animals as described, from partridges and such like. Near to these are the fish in the deep-sea category.'

[8] Compare *De alimentorum facultatibus* 365.7-8 Helmreich: '[The red mullet] has the hardest flesh of all and is sufficiently friable, which shows that at the same time there is not too much stickiness in it or fat.'

Summary of the thinning diet

The thinning diet may be used for the majority of chronic diseases. Drugs are often not needed and it is in general preferable to use milder versions of foods in meals than concentrated versions in drugs. This diet can be helpful for the kidneys (provided there are no stones) and the joints; and also for those with breathing difficulties, enlargements of the spleen and hardening of the liver. It can even alleviate mild or incipient epilepsy. Certain foods 'cut' or 'thin' the humours:[9] taste, smell and touch will often indicate which foods will have this effect. The best foods for this diet are the alliums, namely onions, garlic, leeks; the best media are vinegar and honey. If taken to excess, these foods attack the stomach and cause distress, and the effect can be seen by the way the alliums attack the skin if applied directly.[10] These are the necessary foods for bodies full of thick, sticky, cold humours.[11] The treatise only covers foods. A doctor's advice is needed in order to avoid an imbalance of humours (cf. n. 10). Foods are arranged in the order: vegetables, seeds, meats, roots, fruit and nuts, liquids (honey, wine, milk). The list of foods is fairly comprehensive, with the focus on foods that are part of the 'normal' diet.

Vegetables

Many green herbs are best (after the alliums), especially if grown at altitude and without too much water. Good plants are fresh versions of those which are used as seeds when dried, such as silphium, mustard, thistle, all the *seris* family [probably Alexanders] etc. Less good are mallows, spinach, courgette, cucumber, gourds and plums which are cold, wet and phlegm-producing. Salty and alkaline plants can cut too, as can bitter ones, though these require cooking. Vinegar-honey and wine vinegar enhance these effects, while olive oil neutralises it. Pickled vegetables (and any other kind of pickle) – *hypotrimmata* with vinegar – are by far the best foods for this diet (unless dates are added, as many cooks do).

Seeds

Seeds comprise seeds and cereals, the worst of which is wheat. This plant is nutritious but produces a sticky, thick humour. The preparation of breads and porridges is considered, the worst being those wheat porridges of Asia whose bad effects are alleviated only with yeast,

[9] *Temnein* and *leptunein* are the terms used.

[10] Galen has more to say on these properties of the alliums at *De temp.* 1.661 K (trans. Singer): 'some substances which are eaten for nourishment, when placed in contact with the skin, eat through it and make a wound. Examples are mustard, pickles, garlic, and onion. ... The reasons that all these substances, which cause injury externally, do not do so when taken internally are: that they are changed and transformed by digestion in the stomach and the blood-making process in the veins; that they do not remain in one place, but are divided into small parts which are carried in all different directions; that they are mixed with many humours, as well as with other foods taken at the same time as themselves; that their digestion and excretion are carried out quickly, by which processes that in them which is proper to the nature of the animal is assimilated while the excess sharpness is excreted by stomach, urine and sweat. ... (663) The substance is transformed in digestion, broken down, purified, mixed with many others, divided into many parts, carried elsewhere. ... The condition known as ill humour (*kakochumia*) arises frequently as a result either of bad foods or of some decay or putrefaction in the body itself.'

[11] Compare *De san. tuenda* 4 (6.275 K): vinegar and water and honey mixtures thin raw humours without heating. Capers can also have this effect, if taken with vinegar and honey or vinegar and oil. Raw humours can also be thinned with wine provided it is *leptos* and *kirros* or *leukos.*

salt, wine-honey, grated beet or bird soups. Barley has only a small thinning effect. Bitter vetch, though normally classed as cattle-food, also has a thinning effect. Peas are better than beans, the worst of which are *loboi* which are thick-humoured and phlegm-producing.

Fish and meat

The greatest and most abundant source of food for the thinning diet is provided by rock fish[12] and small mountain birds. Animals living in the hills are drier and better in mixture and their flesh is the least sticky and phlegm-producing. The thinning diet requires no domestic pork since pigs are too idle and consume wet foods, though their extremities are acceptable. People who exercise little are recommended to eat fish, particularly wrasse and bream. If rock varieties are not available, mustard should be added to substitutes. Eels are sticky and create phlegm, as do oysters. Torpedo and ray are acceptable but should be eaten with grated beet or white sauce, leeks and pepper. Goat meat and beef are acceptable, though their flesh is hard and difficult to break down. Hare thickens the blood. Partridges and turtle-doves are acceptable if hung for a day: they generate blood that is nether thick nor thin.

Fruit, honey and wine

Fruit is dangerous, as is most wine. Honey produces a genuinely thin humour. Sweet black wines fill the body with thick blood but thin white wines cut thick humours and cleanse the blood. Sweet wines taken with thinning foods cleanse thick, sticky and phlegmatic humours, in the areas of the chest and lungs. If the problem lies in the liver or spleen, wine is harmful.

Milk

Whey is fine for the thinning diet but other milk products, in particular cheese, are not. Milk thickens. If it must be drunk, then asses' milk is best. Honey should also be added.

Some important predecessors

Galen drew on the long history of dietetics. Important predecessors include the Hippocratic *On Ancient Medicine* (19), in which the thickening and thinning of humours in certain conditions are considered; in the Hippocratic *Regimen II*, the thickening and thinning of the humours do not seem to be at issue,[13] but analogous qualities are attributed to some of Galen's foods. Thus vinegar and honey are linked with phlegmatic and moist conditions (52-53), while garlic is hot and purges, as are leeks, and radish dissolves phlegm with its sharpness (54). Thinning does however seem to be in mind in the Hippocratic *Affections*, where (58), we are told of honey (one of Galen's principal thinning agents) 'honey eaten with other foods nourishes and provides good colour, but on its own it thins rather than

[12] That is, those which inhabit a rocky part of the inshore waters of the sea, as opposed to those that live on sandy bottoms or inhabit the deep sea. This is a category of fish identified by Aristotle and Diocles and Diphilus, among others, and readily identifiable in modern zoology. See, for example, A. C. Campbell, *The Hamlyn Guide to the Flora and Fauna of the Mediterranean Sea* (London 1982). Such species include a number of wrasses and breams.

[13] In his edition, *Hippocrate: Du régime* (Paris 1967) 45, Robert Joly takes '*pachunei*' to mean 'fait grossir', 'fatten' rather than 'thicken'.

strengthens, since it passes through urine and excrement at an above average rate.' The Hellenistic doctor Diphilus of Siphnos, meanwhile, in *On Food for the Sick and Well*,[14] declares that 'the so-called *keris* [a fish] is of soft flesh, good for the bowels, good for the stomach. The juice from it thickens[15] and purges'; further, 'the cuttle-fish, even when boiled, is tender, good to taste and easy to digest. It is also good for the bowels. The juice from it is thinning of the blood and a mover of evacuation when that is obstructed by piles.' 'The *pelamys* is full of nourishment and heavy, diuretic also and difficult to digest. When preserved, like the goby, it is good for the bowels and thinning.'[16] Diphilus also judged courgettes (an unsuitable plant in Galen's regimen) to be made more thinning if eaten with mustard (a suitable plant in Galen's system). As for vinegar, Galen's other prime thinning agent, Heraclides of Tarentum is reported by Athenaeus (*Deipn.* 2.67d-e = fr. 67 Guardasole) to have said in his *Symposium,* 'vinegar causes some things exposed to the air to curdle, and it acts similarly on the contents of the stomach; yet it also dissolves things in the mass, because of course there are different humours mingled within us' (trans. Gulick).

Galen's method: theory and classification

In *De alimentorum facultatibus,* Galen is looking for balance in the humours in the body. He writes in the concluding paragraph (385.13-16 = 6.747 K):

> 'for every difference that I have described in food there is a mid-point. Thus you would find a mean between hard-fleshed and soft-fleshed, which is neither soft-fleshed nor hard-fleshed, and between thinning and thickening and heating and cooling and drying and wetting.'

This is not what is looked for in the thinning diet, which addresses humoral imbalance that may lead to complications. The adverse humours are to be cut or thinned, though it is not clear to what extent it is blood that is to be thinned. Galen has an interesting passage on this in *De alimentorum facultatibus* 364.17-365.2 (quoted in n. 7 above). What Galen appears to be thinning is phlegm and possibly bile, whether or not in combination is unclear. The principal diseases under consideration are the chronic problems mentioned above, while as therapy the thinning diet is a non-invasive approach which is milder than the use of drugs. Drugs themselves, Galen has reminded us, are often the dried and more concentrated forms of efficacious foods.[17] We can see something of the connection between these conditions and humours in the Hippocratic *Affections*:

> 'Arthritis arises from bile and phlegm, when they are set in motion and settle in the joints.' (30)

[14] Quoted by Athenaeus, *Deipn.* 8.355d-356f and (on courgettes) 2.59b.

[15] *pachunei: parugrainei* Gulick.

[16] In these passages, as in the Hippocratic examples, it is not certain that *leptunein* applies to the humours as opposed to the bodily tissues; in the case of the cuttle, however, Diphilus speaks clearly of a 'juice that is thinning of the blood'.

[17] I am grateful to Armelle Debru for the suggestion that Galen might have in mind in this treatise a process depending on the thickness and thinness (or lightness) of particles, such as he describes in *De simp. med. temp. ac fac.*. On this see A. Debru, 'Philosophie et pharmacologie: la dynamique des substances *leptomeres* chez Galien', in A. Debru, ed., *Galen on Pharmacology* (Leiden 1997) 85-102.

'Gout is the most violent of all such conditions of the joints, as well as the most chronic and intractable. In it blood is corrupted by bile and phlegm.' (31)

'Let the phlegmatic patients have their bodies thoroughly dried and made lean by foods, drinks, vomiting, as many exercises as possible, and walks; in spring clean upwards with hellebore.' (20, trans. Potter)

Galen has phlegm in mind in *De subtiliante* 12 (449.33-450.23), where he discusses the danger of clogging the channels to the liver – I presume the thickness or thinness of the humours affects this phenomenon. In such cases, no sweet foods should be consumed. Sweet foods are however unproblematic if thinning is required in the area of respiration, namely the chest and the lungs. Why is Galen concerned with these sweet foods and sweet wines, which could simply be forbidden to patients with these chronic disorders? This brings us to his approach to the diet.

Galen starts with members of the allium family and moves through vegetables, cereals and other plants to fish, birds and meat, thence returning to roots and fruit. He concludes with honey, wine and milk. There is no strict categorisation of food here such as we find in *De alimentorum facultatibus*. Beans take us to sea urchins and shell fish, roots and vegetables return after the meat section. We appear to start with the most efficacious foods, the alliums, and to end with the worst, cheese. But extremely good foods for the thinning diet, such as rock fish and wild birds, are not promoted to the fore, and honey, which has remarkable qualities in this regard, is left to the end, along with the distinctly unpromising wine and milk. We could perhaps set Galen's categories in descending order of suitability as green vegetables (the very best), certain cereals, certain fish and meat, with wine and milk least suitable. There are then repeated references to the excellent honey, along with vinegar: these are the two mainstays, which support the excellent green vegetables. They often make green vegetables even more efficacious and can do much for otherwise unsuitable foods.

There is however an alternative explanation. Marinone claims that there is no good order to the sequence of foods, that the order is not worked out in the way later found in *De alimentorum facultatibus*. Galen makes clear in that treatise that he is there following the traditional order for a dietetic treatise, that is cereals first (as the base of the human diet), followed by beans, vegetables and herbs, land animals and sea animals.[18] But it need not follow, and indeed is unlikely to follow, that Galen has not ordered the foods in our treatise in any way at all. Rather, in our treatise, Galen has adopted another, unorthodox, order.

The order follows the sequence of the traditional Greek meal, seen for example in the *Deipnosophistae* of Athenaeus or the *Hedupatheia* of Archestratus.[19] Vegetables or cereals

[18] In *De alimentorum facultatibus* 264 Galen writes that the writers of *didaskaliai peri ton edesmaton* begin with the seeds of Demeter because the food of wheat is the most useful, and that he has done the same. Plants and animals will have to go into different books. In *De alimentorum facultatibus,* honey cakes come near the beginning, quoting *De subtiliante diaeta* (223) from near its end. In *De alimentorum facultatibus*, the alliums are placed at the end of the plant section in book 2, after roots, from which they were carefully separated in *De subtiliante diaeta*. This reflects order and priority rather than properties or *dunameis*. In *De alimentorum facultatibus*, for example, Galen writes of the alliums in a way that would be quite acceptable in *De subtiliante diaeta*: (329.23-330.4) 'they have a sharp property and accordingly heat the body and thin thick humours in them and cut the sticky ones. The sharpness can be reduced by boiling two or three times but it still thins even then and provides small nourishment to the body.'

[19] For the order of foods treated in Archestratus, see S. D. Olson and A. Sens, *Archestratos of Gela* (Oxford 2000) xxv.

are followed by fish, meat, wine and dessert. Galen then considers the best and the worst in each category. That is why the full range of grains is considered, even though in principle some, such as the range of wheats, are inimical to the thinning diet. Consider, for example, the advice on beans:

> the so-called lobe-beans, too – the cultivated variety – are made into a dough, like that of broad beans. It is important to be aware that this too is thick-humoured and phlegm-producing, although lobe-beans are less flatulent than broad beans and do not have their purgative quality. I need hardly add descriptions of all the other bad seeds, which are universally avoided anyway, without my having to warn against them. Suffice it to say, in summary, that barley is the grain preferable to all other seeds from the point of view of the thinning diet; that pan-baked loaves of wheat have second place; and that one should endeavour to abstain from all others, except perhaps from peas, lentils or spelt, of which one may, if one wishes, partake sparingly. (441.10-19, trans. Singer)

Clearly the characteristics of the foodstuff itself are important, but so too is the mode of preparation. Thus in the quotation above, a *dough* of lobe-beans is particularly undesirable, while wheat loaves *from a pan* are better than some others. So too with barley, it is not envisaged that the cereal will be consumed *tout court*:

> one should try to take barley, too, with honey or with some wine of the same type as Falernian, not with *siraion* or with substances which contain a considerable degree of thickness. (439.30-31, trans. Singer)

There are interesting points of contrast between this treatise and the later *De alimentorum facultatibus*. Here, advice is for clients that doctors expect to treat, and the advice normally explains how the average diet should be modified to take account of the conditions in question. The later treatise aims to provide a comprehensive list of foods and so includes various plants that are normally considered cattle food but which people do eat when starving. The social range of eaters is thus much wider in the later work, with particular interest focussed on peasants in Asia Minor, whose diet based on low-grade cereals provides valuable scientific data. It is worth noting, though, that some of that data is mentioned in passing in the earlier work. Thus, on wheat, he notes:

> Nor should one eat wheat simply cooked – a practice which I have observed to be widespread among peasants in Asia, who season it with a little salt. (438.23-4, trans. Singer)

Similarly, the later classification concerns itself with technical terms for plants, and the difficulty of identification since names vary according to place and to author. This is a scientific problem which, again, Galen mentions in passing in our, earlier, treatise. So, on the identification of the plant, *seris*, he writes:

> there is another kind of 'wild herb' which is less cutting than those mentioned; this kind appears to belong between the two, having neither a definitely cutting nor a thickening effect. The general name for these is *seris*; but the individual species are given different names by rustics, such as lettuce, chicory, the Syrian *gingidia*, and countless similar ones

> in every region. The Athenians use the term *seris* indiscriminately for all of them; for the ancients did not allot any names to the individual species. (435.20-28, trans Singer)

Similarly, the problem arises in the identification of wheats:

> One should not seek any other edible form of wheat-grains apart from these, as far as the Greek-speaking lands are concerned. I cannot be certain to which of the two the ancients gave the name *zeia*, but there is no third kind similar to these two; the name was applied either to one-grain wheat or to *olyra*. (Mnesitheus applies the two terms, 'one-grain wheat' and '*olyra*', to the same seed, regarding *zeia* as a third type inferior to these. His term *zeia* seems to refer to what we now call one-grain wheat, or something even worse than this.) (440.3-11, trans. Singer)

Mnesitheus is the only authority quoted in our treatise, and his deficiencies in classifying wheats are treated at greater length in *De alimentorum facultatibus*, as are the difficulties relating to the development of nomenclature and the desire of the 'Atticists' to return to a notionally purer form of ancient Greek based on the Attic dialect. For the moment, Galen's focus is much more on a certain class of patient than these broader methodological problems.

Galen covers the whole diet, and has to review the cereal staple, wheat, and the meat staple, pork, (as far as affluent clients rather than peasants were concerned) even though, for this diet, they are both eminently unsuitable. He recommends certain foods that in literary texts prompt strong disapproval, such as sauces with vinegar (*hypotrimmata*), fish, tasty pickles, and silphium. All these are beneficial to the thinning diet even though moralists disapprove because they are tasty and therefore foods that promote appetite.[20] He further defines the diet by setting boundaries to what human beings normally eat:

> The horse and the ass would only be consumed by someone who was himself an ass; and to eat leopards, bears or lions you would have to be a wild beast. ... As for dogs and foxes – I have never tasted their meat, since it is not the custom to eat it either in Asia or in Greece, or indeed in Italy. But there are certainly many parts of the world where they are eaten, and my conjecture would be that their effects would be similar to those of the hare; for hare, dog, and fox are all equally dry. (443.15-29, trans. Singer)

It is notable that Italy follows Asia Minor and Greece in the areas under consideration. This may or may not indicate that the treatise was written when Galen was resident in Pergamum. There is a similar preference for Asian and Alexandrian evidence over Italian in *De alimentorum facultatibus*. It is perhaps surprising that Rome is not of more interest in the present work since it is identified as a wet place which prompts severe problems with phlegm in *De temperamentis* 1.630 K.[21] Similarly, more attention is paid to little-known Asian wines than to famous Italian ones. Galen considers some Asian wines unsuitable for the thinning diet because they are insufficiently sweet. He goes on to note:

[20] On foods that promote appetite, see in particular J. Davidson, *Courtesans and Fishcakes* (London 1997).

[21] There is a general preference for animal foods derived from the mountains over those of the plains and of agriculture since the former are drier and also more likely than farm animals to have led an active life. In both circumstances phlegm is less likely to be produced.

> Such wines are not held in high regard, and so people tend to be unaware of their existence, in spite of the fact that they are in fact quite widely produced. They are not profitable wines for the merchants, nor will a person who owns such a wine take much note of it. After all, no one is going to serve a sour, thick, dark wine at a drinking party, at a wedding, at a religious festival, or indeed at any other kind of celebration. (448.20-25, trans. Singer)

Galen seems to imply that these are wines that people drink almost without thinking about it. This is an injurious aspect of the diet, like the simply-prepared wheats mentioned above, which people in need of thinning agents simply cannot ignore.

Conclusion

Galen appears to have written *De subtiliante diaeta* as a short treatise with a particular focus on the thinning of the humours, in the context of the ancient diet. His analysis appears to be broadly in line with that of his predecessors, who would not have been surprised by anything that he had to say on wheat, barley, pork, fish, honey or vinegar. In its focus it is an interesting forerunner of Galen's extensive catalogue of foods in *De alimentorum facultatibus.* It identifies problems that will later be treated at length, in particular regional variation and the correct use of technical terms, and shares with the later treatise a wider social reference than the Hippocratic author had used in *Regimen II.*

University of Exeter

PRE-STOIC HYPOTHETICAL SYLLOGISTIC IN GALEN'S *INSTITUTIO LOGICA*

SUSANNE BOBZIEN

The text of the *Institutio Logica* (*IL*) or *Introduction to Logic* is not found in Kühn because its sole surviving manuscript was first published, not long after its discovery, in 1844, and thus too late for inclusion in Kühn. Moreover, some have thought the work to be spurious.[1] The reasons given for this assumption were on the whole unconvincing. I take it for granted that the *Institutio Logica* is by Galen.

In this paper I trace the evidence in the *Institutio* for a hypothetical syllogistic which predates Stoic propositional logic. It will emerge that Galen is one of our main witnesses for such a theory. In the *Institutio*, Galen draws from a number of different sources and theories. There are the so-called ancient philosophers (οἱ παλαιοὶ τῶν φιλοσόφων); there is the Stoic Chrysippus, whose logic Galen studied in his youth.[2] There are the 'more recent philosophers' (οἱ νεώτεροι), post-Chrysippean Stoics or logicians of other schools who adopted Stoic terminology and theory.[3] There are from the 1st century BC the Stoic Posidonius and the Peripatetic Boethus, both of whom Galen may have counted among the 'more recent philosophers'. Again, in some passages Galen seems to draw from contemporary logical theories of non-Stoic make, presumably of Peripatetic or Platonist origin; and in others he explicitly introduces his own ideas.[4] But apart from Plato, who is generously credited by Galen with the *use* of the later so-called second hypothetical syllogism, the only promising candidates for pre-Stoic proponents of a hypothetical syllogistic are the above-mentioned 'ancient philosophers'. In the following I concentrate on their theory.

There are four passages in the *Institutio* in which Galen mentions the ancient philosophers or, in short, the ancients: one on hypothetical premisses (*IL* 3.3-4), one on hypothetical syllogisms *IL* 14.2), one on epistemology (*IL* 3.2, towards the end), and one on categorical syllogistic (*IL* 7.7).[5] I assume that in the *Institutio*, when Galen says οἱ παλαιοὶ <i.e. τῶν φιλοσόφων> he always has the same philosophers in mind: that is, philosophers of the same period and of the same philosophical persuasion. I assume further that in the *Institutio* the only passages that report the views of the ancients are (i) those in which Galen explicitly

1 E.g. C. Prantl, *Die Geschichte der Logik im Abendlande* vol. I (Leipzig 1855) 591-92.

2 Cf. Galen, *On my own books*, 43 (Kühn xix).

3 Cf. J. S. Kieffer, *Galen's Institutio Logica* (Baltimore 1964) 130-32; J. Barnes, 'Form and Matter', in A. Alberti, ed., *Logica, Mente e Persona* (Florence 1990) 7-119, at 71-23.

4 E.g. in chapters 16-17 of the *Institutio*.

5 However, at the beginning of *IL* 3.2, with τοῖς Ἕλλησιν ... τοῖς παλαιοῖς Galen presumably refers more generally to the Greeks some time before his own.

refers to them and (ii) those in which he uses the same specific terminology as that which he attributes to the ancients. This criterion provides one further passage, *IL* 3.1.

In the following, I first examine the two passages on the hypothetical syllogistic of the ancients in order to establish what this theory was – as far as this is possible from the *Institutio*; then I try to establish who the ancients were by looking at the remaining two *Institutio* passages, and at some other texts by Galen; third, I present some passages in non-Galenic sources that report elements of an early, non-Stoic hypothetical syllogistic and which provide close parallels to the minimal theory of the ancients in the *Institutio*; fourth I discuss a text that elaborates on a theory like that in Galen; and finally I briefly demonstrate how my reconstruction of the hypothetical syllogistic of Galen's ancient philosophers helps to solve some puzzles in the *Institutio*.

1. The theory of the ancients according to the *Institutio Logica*

First then, the *Institutio* passages that attribute elements of a hypothetical syllogistic to the ancients. From them we can collect the basic tenets of the hypothetical syllogistic of these ancients. At *IL* 14.2 Galen writes:

> For such problems[6] we mainly use the hypothetical premisses, which the ancients divided into those in accordance with a connection and those in accordance with a division.[7]

This sentence shows first that the ancients had at least some elementary theory of hypothetical premisses (προτάσεις). (In ancient logic πρότασις can mean either 'premiss' or something like 'proposition', and Galen oscillates between the two.[8] For reasons that will become clearer in the following, I translate πρότασις as 'premiss' where it is part of the theory of the ancients.) Second, *IL* 14.2 shows that the ancients distinguished two basic kinds (εἴδη) of hypothetical premisses and called a premiss of the one kind a 'hypothetical premiss in accordance with a connection' (ὑποθετικὴ πρότασις κατὰ συνέχειαν), and a premiss of the other kind a 'hypothetical premiss in accordance with a division' (ὑποθετικὴ πρότασις κατὰ διαίρεσιν). This is confirmed by *IL* 3.3 (and 3.4, see below), although here the second type of premiss is called a 'dividing' (διαιρετική)[9] hypothetical premiss:

> Now, the ancients called a premiss hypothetical in accordance with a connection mostly in cases when we believe that something is, because something else is, but also in cases

6 The problems mentioned in the passage are whether certain things (fate, providence, gods, the void) *exist* – as opposed to whether things have certain properties.

7 ἐν οἷς προβλήμασι μάλιστα χρώμεθα ταῖς ὑποθετικαῖς προτάσεσιν, ἃς <εἰς τὰς> κατὰ συνέχειαν καὶ κατὰ διαίρεσιν ἔτεμον οἱ παλαιοί (*IL* 14.2).

8 πρότασις could mean 'premiss' e.g. at *IL* 7.1, 7.6 & 7.8, 8.1-3, 9.1-3, although the case is hardly ever clear-cut. (*IL* 8.3, lines 4-6, Kalbfleisch, shows Galen's uncertainty about what logical status a πρότασις has.) Theophrastus seems to have thought that πρότασις has several senses (Alex. *APr.* 11.13-16 ' F (Fortenbaugh) 81A); Alexander was aware of the double meaning of πρότασις as premiss and proposition (Alex. *APr.* 44.17-21).

9 Also at *IL* 3.4. διαιρετική may be an expression introduced later as an abbreviation for κατὰ διαίρεσιν.

> when we think that since something is not, something is, such as <when we think that' since it is not night, it is day.[10] (*IL* 3.3)

(For short I shall sometimes refer to the first type as 'connecting premiss', to the second as 'dividing premiss'.) From the formulations in *IL* 3.3 (τὸ *ἕτερον* εἶδος τῶν προτάσεων), and at *IL* 14.2 (ἃς *<είς τὰς>* κατὰ συνέχειαν καὶ κατὰ διαίρεσιν *ἔτεμον* οἱ παλαιοί) we can infer that the ancients distinguished exactly these two types of hypothetical premiss. At *IL* 3.3 we learn further about the first kind of hypothetical premisses that 'when we believe that something is, because something else is', then the premiss (i.e., I take it, the premiss we use to put forward this belief of ours) is a connecting hypothetical premiss. What is connected here are the 'something' and the 'something else'.

With respect to entire syllogisms, the *Institutio* suggests that the ancients had at least a rudimentary theory of hypothetical syllogisms, or more precisely, of syllogisms that come to be from hypothetical premisses. For the above-quoted passage at *IL* 14.2 continues thus:

> The Stoics call the connecting hypothetical premisses conditional assertibles, and the dividing ones disjunctive assertibles, and they (i.e. the ancients and the Stoics) agree that two syllogisms come to be with the conditional assertible, and two with the disjunctive.[11]

(The ancients would of course have said 'two ... with the hypothetical premiss κατὰ συνέχειαν, and two with the hypothetical premiss κατὰ διαίρεσιν.) Thus we can assume that the ancients held that two kinds of hypothetical syllogism come to be from connecting hypothetical premisses and two from dividing ones. Again, we can infer that the ancients distinguished exactly four basic kinds of hypothetical syllogisms (*IL* 14.2): as they had two types of hypothetical premisses and each provides two kinds of hypothetical syllogisms, we end up with four such kinds.[12]

The pair of expressions κατὰ συνέχειαν and κατὰ διαίρεσιν appears to be a distinguishing mark of the hypothetical syllogistic of the ancients: this terminology is sufficiently rare in ancient logic – in fact, the pair together never occur anywhere else in ancient logic,[13] and Galen never ascribes it to anyone else. Assuming that the expressions are a distinguishing mark, we can add one further section, *IL* 3.1, as a passage in which Galen draws on the theory of the ancients.[14] This section provides us with the following *general*

10 μάλιστα μὲν οὖν ἐπειδὰν ὑπάρχον τι πιστεύηται δι' ἕτερον ὑπαρχειν [ἤ] κατὰ συνέχειαν, ὑποθετικὴ πρὸς τῶν παλαιῶν φιλοσόφων ὀνομάζεται <ἡ> πρότασις, ἤδη δὲ καὶ ἐπειδὰν [μὲντοι] διότι μὴ ἔστι τόδε, εἶναι τόδε νοῶμεν, οἷον <διότι> νὺξ οὐκ ἔστιν, ἡμέραν εἶναι· μάλιστα μὲν οὖν ὀνομάζουσι τῶν τοιαύτην πρότασιν διαιρετικήν (*IL* 3.3). del. ἤ Prantl; add. διότι Barnes et al.

11 καλοῦσι δὲ τὰς μὲν κατὰ συνέχειαν οἱ Στωικοὶ συνημμὲνα ἀξιώματα, τὰς δὲ κατὰ διαίρεσιν διεζευγμένα, καὶ συμφωνεῖταί γε αὐτοῖς δύο μὲν γίγνεσθαι συλλογισμοὺς κατὰ τὸ συνημμὲνον ἀξίωμα, δύο δὲ κατὰ τὸ διεζευγμὲνον (*IL* 14.2).

12 It has been suggested that συμφωνεῖταί γε αὐτοῖς at *IL* 14.2 could be translated as '... and it is agreed that among them (i.e. among the hypothetical syllogisms) there are ...' or as '... it is agreed among them (i.e. the Stoics) that there are'. But the first would be a rather unusual reading of the Greek, and the second would make Galen say something that seems entirely unmotivated, since in the context of the sentence disagreements among the Stoics are not at issue at all. If either of these readings *was* what Galen intended, plainly we would have no evidence that the ancients had exactly four types of hypothetical syllogisms.

13 There are remnants and modifications of this terminology in several later Peripatetic and Platonist texts, but we do not find exactly this pair of expressions in any one text other than Galen.

14 The expression κατὰ συνέχειαν occurs also at *IL* 5.5, but the rest of that passage is undoubtedly Stoic.

account of the hypothetical premisses: Hypothetical premisses are those 'in which we make the statement ... about what is if something is, and <those in which we make the statement about> what is if something is not.' Then Galen adds separate descriptions of the two kinds of premisses: the hypothetical premisses are called connecting 'when if something is, necessarily something else is'; and they are called dividing 'when either if something is not, something is or if something is, something is not'.[15]

Galen's *general* account is conjunctive. I assume that the first conjunct characterizes the connecting hypothetical premisses, and thus corresponds to their specific description, and that this in turn corresponds to the above-mentioned account of the connecting premiss of the ancients given at *IL* 3.3.[16] Thus the connecting hypothetical premiss is explicated by the clauses

τὴν ἀπόφα[ν]σιν ποιούμεθα περὶ *τοῦ τίνος ὄντος τί ἐστι* and
τινὸς ἑτέρου ὄντος ἐξ ἀνάγκης[17] *εἶναι* λέγωσι *τόδε τι* and
επειδὰν *ὑπάρχον τι* πιστεύηται *δι' ἕτερον ὑπάρχειν.*

Equally, I assume that the second conjunct of the general account characterizes the dividing hypothetical premisses, and that it corresponds to their specific account. Thus the dividing hypothetical premiss is explicated by the clauses

<τῶν ἀπόφανσιν ποιούμεθα περὶ τοῦ> τίνος οὐκ ὄντος τί ἐστιν and
ἤτοι μὴ ὄντος εἶναι ἢ [μὴ] ὄντος μὴ εἶναι <λέγωσι τόδε τι>.

Note that in all cases here the hypothetical premisses are characterized neither by a certain linguistic form (e.g. 'if p, q'), nor by the use of certain connective particles (such as 'if' or 'and'), nor as being a combination of simple propositions (e.g. 'p' and 'q' in 'if p, q') – although such characterizations are common in Stoic logic. Rather, the hypothetical premisses are defined and classified with respect to the sort of things *about* which *in* them a statement or assertion is made:[18] In hypothetical premisses characteristically a statement is made about a relation[19] between 'things' (πράγματα).[20] (The things are presumably either something like states of affairs such as that humans are animals, or generic 'things' such as human and animal, see below.) These relations are, I assume, the binary relations of

15 Γένος ἄλλο προτάσεώς ἐστιν ἐν αἷς τῶν ἀπόφα[ν]σιν οὐ περὶ τὰς ὑπάρξεως ποιούμεθα τῶν πραγμάτων, ἀλλὰ περὶ τοῦ τίνος ὄντος τί ἐστι καὶ τίνος οὐκ ὄντος τί ἐστιν· ὑποθετικαὶ δὲ ὀνομαζέσθωσαν αἱ τοιαῦται προτάσεις, αἱ μὲν, ὅταν τινὸς ἑτέρου ὄντος ἐξ ἀνάγκης εἶναι λέγωσι τόδε τι, κατὰ συνέχειαν, αἱ δὲ, ὅταν ἤτοι μὴ ὄντος εἶναι ἢ [μὴ] ὄντος μὴ εἶναι, διαιρετικαί (*IL* 3.1). προτάσεως ms, Barnes et. al., προτάσεων Mynas, Kalbfleisch; del. μὴ Prantl.

16 The difference is that one (*IL* 3.1.) uses εἶναι, the other (*IL* 3.3) ὑπάρχειν, and Galen has in between, at *IL* 3.2, just explained that the two verbs were used synonymously in this context.

17 The use of a phrase in the description of a connecting hypothetical premiss that indicates necessity (presumably *necessitas consequentiae*) may be another mark of early Peripatetic theory, cf. e.g. Arist. *APr.* I 32 47a28-31, *Top.* B4 111b19, and my 'Wholly hypothetical syllogisms', *Phronesis* 45 (2000) 87-137, at 92-23 and 113.

18 Moreover, at *IL* 3.3 they are explained with respect to what they 'say', and with respect to the situations in which we use them (i.e. when we have what sort of beliefs).

19 Amm *Int.* 74.2-3 refers to these relations as σχέσεις.

20 By contrast, in categorical premisses we make a statement about (the ὕπαρξις, i.e. holding or not holding of) the things (*IL* 3.1, cf. Amm. *Int.* 4.7-11).

connection (συνέχεια) and division (διαίρεσις) 'in accordance with which' the premisses come about (*IL* 3.3, κατὰ συνέχειαν γίγνεσθαι).

Whenever either a connection or a division holds between two things, this fact can be described in terms of a dependency that exists between these things or their being or not being (holding or not holding). This is clear from Galen's formulations, i.e. from his use of a participle construction (for that on which something depends) and a main clause (for that which depends on it).[21] Thus, in the case of a connection, there is a dependency of one thing's being on another thing's being. This suggests a relation that is not symmetrical. In the case of a division, there is a dependency between one thing's being and another thing's not being. This can be put either as saying that one thing's being depends on another thing's not being, or as saying that one thing's not being depends on another thing's being (*IL* 3.1). This suggests a symmetrical relation.

The relation of dependency which the connecting and the dividing hypothetical premisses share could be described by the formula 'φ ὄντος, ψ ἐστι', where for φ and ψ the things (πράγματα) are put in, either taken affirmatively, or taken negatively – i.e. either *A* or '*A* οὐκ / μή', etc. This formula 'φ ὄντος, ψ ἐστι' indicates a conditionality of sorts, and I believe that it is this conditionality which was captured by the ancient in the expression 'hypothetical' (ὑποθετικός). Note also that the passages leave no doubt that for the ancients the things put in for φ and ψ must differ from each other: recall e.g. the formulation at *IL* 3.1 'if something is, necessarily something *else* is'.

So far the accounts of the hypothetical premisses are vague and unfamiliar to the modern logician. We reach more familiar ground when we look at Galen's examples at *IL* 3.4:

> Thus a sentence such as 'if it is day, the sun is above the earth' is called a conditional assertible by the more recent philosophers, and a connecting hypothetical premiss by the ancients; and sentences such as 'either it is day or it is night' are called a disjunctive assertible by the more recent philosophers, and a dividing hypothetical by the ancients.[22] (*IL* 3.4)

This passage makes it clear that – at least by Galen – the hypothetical premisses were considered as λόγοι, sentences, and thus as linguistic items. We can also infer from the examples that the connecting hypothetical premisses of the ancients must have had some kind of conditional form, and the dividing ones some kind of disjunctive form, and that they must be at least superficially comparable with the Stoic conditionals and disjunctions.[23]

This information, together with the above-quoted passage from *IL* 14.2 gives us a very rough picture of the four types of hypothetical syllogism of the ancients. They had two types of hypothetical syllogisms with connecting hypothetical premiss, and two types with dividing

21 In fact, in our passages the main clause is an infinitival construction, depending on a verb of saying.

22 ὡς ὀνομάζεσθαι τὸν μὲν τοιοῦτον λόγον 'εἰ ἡμέρα ἐστίν, ὁ ἥλιος ὑπὲρ γῆς ἐστιν' συνημμένον ἀξίωμα κατὰ γ' τοὺς νεωτέρους φιλοσόφους, κατὰ μέντοι τοὺς' παλαιοὺς πρότασιν ὑποθετικὴν κατὰ συνέχειαν· τοὺς δέ γε τοιούτους 'ἤτοι γ' ἡμέρα ἐστιν ἢ νύξ ἐστι' διεζευγμένον μὲν ἀξίωμα παρὰ τοῖς νεωτέροις φιλοσόφοις, πρότασιν δὲ ὑποθετικὴν κατὰ διαίρεσιν παρὰ τοῖς παλαιοῖς (*IL* 3.4).

23 We cannot infer that the ancients themselves used the examples Galen provides, nor do we have any particular reason to assume they were theirs. For all we know, they could have been 'term logical', as may be suggested by Boethius, who in his *De Hypotheticis Syllogismis* seems to present a theory very similar to that of 'the ancients' in Galen (see below, Section 5), and also by the way Galen talks about conversion in a hypothetical πρότασις at *IL* 6.4.

hypothetical premiss (see above). We may now assume that the first two hypothetical syllogisms of the ancients were such that they could be conceived of as being similar to the first two Stoic indemonstrables:

(1)	If p, then q *Now p* Hence q	(Α ὄντος Β ἐστι)	(2)	If p, then q *Now not q* Hence not p	(Α ὄντος Β ἐστι)

And the other two kinds of hypothetical syllogisms should be somewhat similar to the fourth and fifth Stoic indemonstrables:

(4)	Either p or q *Now p* Hence not q	(Α μὴ ὄντος Β ἐστι)	(5)	Either p or q *Now not p* Hence q	(Α ὄντος Β μὴ ἐστι)

On the above assumption that the hypothetical syllogistic of the ancients comprises only those passages which Galen directly ascribes to them and those with the same distinctive terminology, it seems then that the features that are unique to this theory are the following:

1. There are exactly two main types of hypothetical premisses – as opposed to three or more. In particular no hypothetical premisses are (or have the linguistic form of) negated conjunctions, whereas both the Stoics and later Peripatetics and Platonists had syllogisms with a negated conjunction as main premiss.
2. The two types are called hypothetical premisses in accordance with a connection (κατὰ συνέχειαν) and hypothetical premisses in accordance with a division (κατὰ διαίρεσιν).
3. The relation of division (διαίρεσις) is considered as binary; the dividing premisses always have precisely two disjuncts.
4. Exactly four basic types of hypothetical syllogisms can be constructed from the hypothetical premisses, two with each type.
5. Each basic hypothetical syllogism has one hypothetical premiss, and one that is not hypothetical (and thus presumably categorical); otherwise the identification of these syllogisms with four of the Stoic five types of indemonstrable arguments would become quite incomprehensible.

There is however in the relevant passages of the *Institutio* nothing that tells us whether the hypothetical premisses and syllogisms of the ancients had a specific linguistic form or contained specific expressions, or what the 'things' were that are connected or divided in the hypothetical premisses. For instance, in the case of the first hypothetical syllogism, with the premisses we make statements about that if this is, that is, and about that this is, and with the conclusion we make a statement about that that is. If 'this' and 'that' are something like states of affairs, and 'is' means something like 'obtains', 'is the case', 'is true' or 'holds', the following could be a first hypothetical syllogism:

If *it/this is a human being*, then *it/this is rational.*[24]	If p, then q.
Now *it/this is a human being*.	Now p.
Therefore *it/this is rational*	Therefore q.

24 Here the first 'it' would be demonstrative and the remaining ones either demonstrative or anaphoric.

If 'this' and 'that' are rather terms (in the Aristotelian sense), and 'is/holds' means something like 'has application/holds of something', the following could be a typical first hypothetical syllogism:[25]

If *human being* is/holds (of it/this), then *rational* is/holds.	If A is/holds (i.e. of it/this), then B is/holds.
Now *human being* is/holds.	Now A is/holds.
Therefore *rational* is/holds.	Therefore B is/holds.

This could perhaps also be expressed as:

If it/this is *a human being*, then it/this is *rational*.	If this is A, then it/this is B.
Now it/this is *a human being*.	Now this is A.
Therefore it/this is *rational*.	Therefore this is B.

Alternatively, the following could be a typical first hypothetical syllogism:

If A is/holds (i.e. of something), B is/holds (of it).	If *human being* is/holds (of something), then *rational* is/holds (of it).
Now A is/holds (of this).	Now *human being* is/holds (of this).
Therefore B is/holds (of this).	Therefore *rational* is/holds (of this).

This could perhaps also be expressed as:

If something is A, it is B.	If something is *a human being*, then it is *rational*.
Now this is A.	Now it/this is a *human being*.
Therefore this is B.	Therefore it/this is *rational*.

Similar possibilities can be conceived of for the other three types of hypothetical syllogism. The general uncertainty about what the ancient philosophers assumed to be the logical structure of their hypothetical syllogisms is frustrating. I have dwelt on this point, since I believe that it is important for our understanding of the development of a Peripatetic and Platonist hypothetical syllogistic that we make no rash assumptions about their original logical form; in particular, that we do not simply assume that they were understood to have the forms the Stoics later gave their indemonstrables, and which would be fitting for a propositional logic.

25 If this possibility surprises you, compare Ammonius' introduction of hypothetical προτάσεις: ὑποθετικοῦ (i.e. ἀποφαντικοῦ λόγου) δὲ τοῦ σημαίνοντος τίνος ὄντος τὸ ἐστιν ἢ οὐκ ἔστιν, ἢ τίνος μὴ ὄντος τί ἐστιν ἢ οὐκ ἐστιν, ὡς ὅταν εἴπωμεν 'εἰ ἄνθρωπός ἐστι, καὶ ζῷόν ἐστιν, εἰ ἄνθρωπός ἐστι, λίθος οὐκ ἔστιν, εἰ μὴ ἔστιν ἡμέρα, νύξ ἐστιν, εἰ μὴ ἔστιν ἡμέρα, οὐκ ἐστιν ἥλιος ὑπὲρ γῆν.' 'The hypothetical (i.e. assertoric sentence) indicates what is or is not if something is, or what is or is not if something is not, as when we say "if it is a human being, it is an animal", "if it is a human being, it is not a stone", "if it is not day, it is night", "if it is day, the sun is not above the earth".' Here at least the first two examples, which are clearly Aristotelian, suggest a term-logical understanding of the hypothetical assertoric sentences.

2. Who were the παλαιοί? The evidence in Galen

I now turn to the question who the ancient philosophers were. In the present section, I consider the passages in the *Institutio* and in other Galenic writings. In the *Institutio* the ancients are contrasted in ch. 14 with the Stoics and in ch. 3 with 'the more recent philosophers', who use Stoic terminology. We can infer that the ancient philosophers (i) were not Stoics, (ii) were older than those Stoics, (iii) were older than those more recent philosophers.

Furthermore, we have some good reasons for assuming that the ancient philosophers were Peripatetics: first, the terminology they use is Peripatetic: e.g. πρότασις, 'hypothetical premiss', 'hypothetical syllogism';[26] second, the account of the hypothetical premisses in terms of things (πράγματα) is Peripatetic;[27] third, the predicate of holding (ὑπάρχειν) is used not of the truth-bearers (as the Stoics would do), but of those things;[28] fourth, the two things that are related in a hypothetical premiss must differ from each other, whereas for the Stoics they could be the same;[29] fifth, at *IL* 7.7 the ancient philosophers are said to distinguish three figures of categorical syllogisms, and these syllogisms and their classification are without doubt Peripatetic.

In addition, everything suggests that the ancient philosophers were *early* Peripatetics, i.e. Aristotle, Theophrastus, Eudemus, and their contemporaries.[30] And in the case of hypothetical syllogistic, since as far as we know Aristotle did not have such a thing,[31] the ancients could only have been Theophrastus and Eudemus and their contemporaries. Thus in *On the doctrines of Hippocrates and Plato* 2.2.4, Galen says: 'the ancient philosophers (who are associated with) Theophrastus and Aristotle'[32] and in his *On the Method of Treatment* 1.3 we read that Theophrastus was well-practised in logic.[33] Galen also tells us that he wrote a commentary in six books on Theophrastus' *On affirmation and negation* (περὶ καταφάσεως καὶ ἀποφάσεως), and three books on Eudemus' *On Speech* (περὶ λέξεως).[34] In the *Institutio* itself, at *IL* 3.2, Galen digresses briefly from his topic of propositional logic and introduces some epistemological terminology. Here, in the context of the difference between νοήσεις and ἔννοιαι, Galen says: 'there are also other (i.e. non-empirical) concepts, ... which are innate to all human beings; the ancient philosophers call these when they are expressed by

26 Cf. Philop. *APr.* 242.24-243.10, and my 'Stoic Hypotheses and Hypothetical Argument', *Phronesis* 1997, 299-312. The expressions ἀπόφα[ν]σις, at *IL* 3.1, and πρόβλημα, at *IL* 14.2 also suggest a Peripatetic origin.

27 Cf. Philop. *APr.* 242.27-8 (= F 111B), where, after having mentioned Theophrastus and Eudemus in the previous paragraph, Philoponus states that the Peripatetics call the things (πράγματα) things (πράγματα); cf. also [Amm.] *APr* 68.4-5.

28 Cf. Alex. *APr.* 156.29-157.2 (= F 100B) for the use of ὑπάρχειν by Theophrastus.

29 See above, pp. 61-62.

30 Theophrastus and Eudemus wrote books entitled Περὶ ἑρμηνείας, and 'Αναλυτικά (Philop. *Cat.* 7.20-22), and in particular Theophrastus wrote many more books on logic (Diog. Laert. 5.42 and 50). Of their contemporaries, we know that Phaenias of Eresus and Strato of Lampsacus wrote on logic. The first seems to have written works entitled Κατηγορίαι, Περὶ ἑρμηνείας, and 'Αναλυτικά (Philop. *Cat.* 7.20-22, Wehrli frg. 8), the second wrote, among other things, an introduction to the *Topics* and a treatise Περὶ τοῦ μᾶλλον καὶ ἧττον (Diog. Laert. 5.59-60).

31 Cf. also Amm. *Int.* 3.15-4.4, [Amm.] *APr.* 67.41-68.4.

32 παλαιοὶ φιλόσοφοι οἱ περὶ Ωεόφραστόν τε καὶ 'Αριστοτέλην, *De placitis Hippocratis et Platonis* 2.2.4 ('F 114).

33 *De methodo medendi* 1.3 (= F 130).

34 Galen, *On his own works* (*De libris propriis*) 11, 14: XIX 42, 47 K.

means of the spoken word, ἀξίωματα'.[35] This is clearly not the Stoic, and most probably a Peripatetic use of ἀξίωμα.[36] In particular, this account is similar to Theophrastus' account of ἀξίωμα, as preserved by Themistius *APost* 7.3-6 (= F 115), where we read: 'Theophrastus defines ἀξίωμα as follows: ἀξίωμα is a sort of belief (either about homogeneous matters ... or about absolutely everything ...) for these are as it were innate and common to all'.[37] Thus the passages on the ancient philosophers in the *Institutio* and in other Galenic writings suggest that they were early Peripatetics and that they included Theophrastus.

3. Early Peripatetic hypothetical syllogistic according to other sources

Next, I look at our sources for early Peripatetic hypothetical syllogistic (other than the *Institutio*), in order to show that the surviving evidence of Theophrastus' and Eudemus' theories tallies amazingly well with the theory of the ancient philosophers as given in Galen's *Institutio*.

First, there are several texts that report that Theophrastus and Eudemus discussed hypothetical *syllogisms*; moreover, Theophrastus and Eudemus are the earliest philosophers for whom there is such evidence.[38] This certainly makes them good candidates for being Galen's ancient philosophers.

Second, in a passage in which he comments on Aristotle's remarks on 'the other syllogisms from an hypothesis' at *Prior Analytics* 41a37, and in which he mixes Stoic, Aristotelian, and Peripatetic terminology, Alexander of Aphrodisias (*APr.* 262.28-32 = F 112A) reports about the 'old' philosophers (οἱ ἀρχαίοι) that they said that Aristotle's 'syllogisms from some other hypothesis' (which Alexander also calls hypothetical syllogisms) were mixed from a hypothetical premiss and a probative (that is categorical) premiss.[39] Alexander identifies the hypothetical syllogisms of the 'old philosophers' with the Stoic types of indemonstrables.[40] Theophrastus is the only philosopher mentioned by name in the immediate context of the

35 τοιαύται (i.e. the ἔννοιαι) δ' εἰσὶ καὶ ἄλλαι ... ἀλλ' ἔμφυτοι πᾶσιν ὑπάρχουσαι ... .

36 Cf. Aristotle *Top.* A 172a27-30, *Met.* B.1 995b8, B.2 996b27-9, G3, 1005a24. The Stoics use the term ἀξίωμα in order to denote propositions, a fact Galen acknowledges shortly after, at *IL* 3.3.

37 ὁ γὰρ Θεόφραστος οὕτως ὁρίζεται τὸ ἀξίωμα· ἀξίωμά ἐστι δόξα τις ἡ μὲν ... ἡ δὲ ἄπλῶς ἐν ἅπασιν, ... ταῦτα γὰρ καθάπερ σύμφυτα καὶ κοινὰ πᾶσι.

38 The texts are Alex. *APr.* 389.31-390.3, Boethius *HS* 1.1.3-4, Philop. *APr* 242.14-20, and Al-Farabi *Int.* 53.6-12 (Kutsch and Marrow) (= F 111E, 111A-C). There is disagreement in the sources about the extent of this discussion, Boethius claiming that Theophrastus and Eudemus dealt with hypothetical syllogisms only briefly, Philoponus that they did so at length. I do not believe that Theophrastus and Eudemus wrote lengthy treatises on hypothetical syllogisms. For we would expect Alexander, who is our earliest surviving and main source on Peripatetic syllogistic, to have commented on this fact, but he does not. He only reports that Theophrastus and Eudemus mention Aristotle's 'other syllogisms from a hypothesis' (Alex. *APr.* 390.2). Besides, Boethius seems more reliable than Philoponus on this point, since – unlike the latter – he distinguished between Eudemus' and Theophrastus' views. (There is no evidence that any early Peripatetics discussed hypothetical προτάσεις separately from the hypothetical syllogisms. This squares well with my assumption that they considered hypothetical προτάσεις only in the context of hypothetical syllogisms, and thus as hypothetical premisses, and not independently as propositions.)

39 οὓς οἱ ἀρχαίοι λέγουσι μικτοὺς ἐξ ὑποθετικῆς προτάσεως καὶ δεικτικῆς, τουτ' ἐστι κατηγορικῆς. Here πρότασις should mean 'premiss', since the two kinds of πρότασις have their names according to their *function* in this Aristotelian type of a syllogism from a hypothesis. I have argued this point in detail elsewhere ('The development of *modus ponens* arguments in antiquity', *Phronesis* (2002, forthcoming).

40 Alex. *APr.* 262.28-32.

Alexander passage, and he is mentioned in the context of an interpretation of Aristotle's syllogisms from a hypothesis (Alex. *APr* 263.10-14). Theophrastus is hence most probably one of the 'old philosophers' of the passage, and most probably the old philosophers were elaborating on Aristotle's syllogisms from a hypothesis.[41]

The information in this passage about the old philosophers squares very well with that in Galen about the ancients. They must be Peripatetic philosophers. They discussed hypothetical syllogisms. These syllogisms had one hypothetical premiss, and one probative (that is, in Alexander's and Galen's terminology, categorical) premiss. They are compared by Alexander to the Stoic types of indemonstrable syllogisms, just as the hypothetical syllogisms of the ancients are by Galen (*IL* 14.2). Moreover, in the passage, as elsewhere, Alexander refers to the Stoic conditional premisses as 'the connecting one' (τὸ συνεχές, Alex. *APr*. 262.33, 263.6 and 22), and to the disjunctive premisses as 'the dividing one' (τὸ διαιρετικόν, Alex. *APr*. 264.7 and 10). As these terms are neither Stoic nor Aristotle's, they should have their origin in, or be derived from, the nomenclature of the third party mentioned in the passage, the 'old philosophers', Theophrastus and consorts. And they are nicely correlated to the terms κατὰ συνέχεια and κατὰ διαίρεσιν of Galen's ancients.[42]

These striking parallels between Galen and Alexander point to the conclusion that Alexander's old philosophers (ἀρχαίοι) in this passage are the same as Galen's ancient philosophers (παλαιοί) in the *Institutio*; that the theory of these old and ancient philosophers took its origin from Aristotle's syllogisms from a hypothesis; that they consisted of one hypothetical premiss and one categorical premiss; and that the latter was called 'probative' by them; furthermore, that the hypothetical premisses were either connecting or dividing and that the latter contained two disjuncts; that they were sufficiently similar to four of the five types of Stoic indemonstrables that later philosophers could identify them with these.

A third further piece of evidence comes from Boethius' *On hypothetical syllogisms* (*HS*) and is about Eudemus:

> Eudemus holds that the hypothesis from which the hypothetical syllogisms obtain their name is said in two ways: for either (a) through a hypothesis of things consistent in themselves something which can in no way happen is accepted in such a way that the argument leads toward the end (destruction?) of the thing; or (b) the consequence which is posited in the hypothesis is revealed (established?, indicated?) by virtue of a connection or by virtue of a division.[43] (Boeth. *HS* 1.2.5, p. 212 Obertello)

I am far from sure how the (b)-clause of this sentence should be rendered (we have to imagine Boethius translating from the Greek), but in any case the passage is important for several reasons:

41 E.g. Theophrastus in *his Analytics*, see Alex. *APr*. 389.31-390.3.

42 I believe that they are Alexander's (or a predecessor's) coinage of two Peripatetic terms for the two types of hypothetical premisses (i.e. *linguistic* items), which he derived from the early Peripatetic terms for the *ontic* relations Galen talks about.

43 Hypothesis namque unde hypothetici syllogismi accepere vocabulum duobus (ut Eudemo placet) dicitur modis. aut enim (a) tale adquiescitur aliquid per quamdam inter se consentientium conditionem, quod fieri nullo modo possit, ut ad suum terminum ratio perducatur; aut (b) in conditione posita consequentia vi coniunctionis vel disiunctionis ostenditur. (Boeth. *HS* 1.2.5, p. 212 Obertello)

(i) Eudemus says of the (kind of) hypothesis which is 'said in two ways', that it is it from which the hypothetical syllogisms get their name. This implies that Eudemus thought that such a hypothesis was commonly part of a hypothetical syllogism. The context of the Boethius quote (Boeth. *HS* 1.2.6 and 7) suggests that the origin of the Eudemian hypotheses of clause (a) are (the hypotheses from) Aristotle's *reductiones ad impossibile*,[44] and that the origin of the Eudemian hypotheses of clause (b) are (the hypotheses from) Aristotle's 'other syllogisms from a hypothesis', which Aristotle mentions e.g. at *Prior Analytics* 41a38 and 50a16-28. Hence I assume that the Eudemian hypotheses both of clause (a) and (b) were so-called because they were understood as *premisses* in *hypothetical* syllogisms – in contrast to being a special type of proposition; and that the ones of clause (b) are those items that are (possibly by Theophrastus, possibly later) also called hypothetical premisses;[45] they are then functionally the same sort of items as the hypothetical premisses of the ancients in Galen's *Institutio*.

(ii) The Boethius passage tells us that Eudemus distinguished two types of hypotheses of his second kind: those in which a consequence is indicated or established by a *coniunctio*, and those in which a consequence is indicated or established by a *disiunctio*. With the pair of expressions *coniunctio / disiunctio* Boethius could translate either the pair of expressions for ontic relations, συνέχεια and διαίρεσις by virtue of which the consequence is *established/revealed* (in which case we have the same distinction as that of the ancients in Galen's *Institutio*);[46] or he could render a pair of expressions for types of hypothetical premisses, such as συνημμένον and διεζευγμένον,[47] in which case these hypotheses *indicate* the consequence.[48] Either way, it is clear that the hypotheses themselves are linguistic items. The parallel to Galen's connecting and dividing hypothetical premisses is obvious.

(iii) Perhaps most interesting is Eudemus' point that for every hypothesis of this second Eudemian type (i.e. from clause (b)) there is a *consequentia*, a consequence, and that this consequence is indicated both in conditional *and in disjunctive* premisses. (Or, that such a consequence can be established either by a connection *or by a division*.) I assume (a) that this consequence is the relation of dependency that is described in the formulations of the kind 'φ ὄντος, ψ ἐστι' in Galen;[49] (b) that this consequence is what gives rise to the expression 'hypothesis' or 'hypothetical premiss';[50] and (c) that it is this consequence which justifies the inference of the conclusion from the premisses in a hypothetical syllogism. (There is a parallel to this point in Galen's *Institutio* at *IL* 14.10, where Galen says that 'the syllogisms that come to be from hypothetical premisses are completed in accordance with a transition

44 These are one kind of the syllogisms called 'from a hypothesis', cf. Arist. *APr.* 40b25-6.

45 Cf. the Alexander passage just discussed.

46 See also Philoponus (*APr.* 245.6, cf. 10), who calls the disjunctive premiss of a hypothetical syllogism a *ὑπόθεσις* κατὰ διαίρεσιν. This draws a direct connection from Eudemus' hypotheses to the hypothetical premisses Galen's ancients.

47 Or τὸ συνεχές and τὸ διαιρετικόν.

48 *ostenditur* could be a translation of σημαίνειν or ἐμφαίνεσθαι, meaning 'indicated'; this would provide a parallel to Amm. *Int.* 3.32-4.1, 73.30-2, 74.2-3.

49 This is perhaps corroborated by the fact that Boethius, when explicating Eudemus' second type of hypothesis, says '<propositiones> quae vero a simplicibus differunt illae sunt, quando aliquid dicitur esse vel non esse, si quid vel fuerit vel non fuerit' (*HS* 1.2.7).

50 Amm. *Int.* 74.2-5 draws this connection explicitly.

(μετάβασις) from one thing to another by means of an ἀκολουθία or a μάχη. Although Galen uses his own terminology (ἀκολουθία and μάχη instead of συνέχεια and διαίρεσις), the restriction to two things ('one thing' / 'another') in the hypothetical premisses points to the ancients as originators of this transition-theory.[51] The 'transition from one thing to another by means of an ἀκολουθία or a μάχη' could be the same thing as Eudemus' '*consequentia* which is posited in the hypothesis <and> is established/revealed/ indicated by virtue of a connection or by virtue of a division'. (However, this is conjecture only.)

As a result of this survey of non-Galenic sources on early Peripatetic (not wholly) hypothetical syllogisms, we can state that in them we find on the one hand parallels for virtually everything Galen attributes to the ancient philosophers and on the other nothing that is incompatible with their position. We have thus every reason to assume that these sources report parts from the same general theory.

Let me illustrate the resulting theory of hypothetical syllogisms of the ancient philosophers as it presents itself now. Take a later so-called first hypothetical syllogism, with a linguistic form of the kinds suggested above in Section 1.

If A (is/holds), B (is/holds).	hypothetical premiss (ὑποθετικὴ πρότασις)
Now A (is/holds).	probative premiss (δεικτικὴ πρότασις)
Therefore B (is/holds)	conclusion (συμπέρασμα)

Here the hypothetical premiss is connecting (κατὰ συνέχεια), the connection (συνέχεια) being intended to hold between the things A and B. There is indicated to be a dependency of B on A, which can also be described as 'A ὄντος, B ἐστι'; and it is this indicated dependency which makes the premiss hypothetical. Once A('s being) has been proved, the dependency allows the transition from A('s being) to B('s being), or the inference of B('s being). The case of the later so-called second hypothetical syllogism is perhaps a little more complex. Take

If A (is/holds), B (is/holds).	hypothetical premiss
Now B (is/does) not (hold).	probative premiss
Therefore A (is/does) not (hold).	conclusion

Here, again, a connection (συνέχεια) is intended to hold between the things A and B, and there is indicated to be a dependency between A and B that can be described as 'A ὄντος, B ἐστι'. This dependency is now meant to allow the transition from B's not being to A's not being (or from 'not B' to 'not A'). But how? There are two possibilities. Either this was just assumed to be an obvious power of the connecting hypothetical premisses. This could have been justified with reference to Aristotle's *Topics* 111b20-23, where Aristotle argues that if one wants to refute something, then one must examine what it is that is if the point at issue is, since then, when we prove that what follows from the point at issue is not, we will have rejected the point at issue. Alternatively, the early Peripatetics believed that the second

51 Galen himself, like the later Peripatetics and Platonists, allows for a plurality of disjuncts in hypothetical propositions.

hypothetical syllogism was in need of prove. We know that this is what Galen thought.[52] This proof could have been the reduction to a first hypothetical syllogism by means of contraposition (ἀντίστροφη) in arguments, i.e. in this case by replacing the second premiss by the contradictory of the conclusion, and the conclusion by the contradictory of the second premiss. Galen describes this type of contraposition at *IL* 6.5, and as Aristotle recognized this kind of reduction, we can assume that the early Peripatetics were familiar with it, too.[53] I do not know which of these two views of the second hypothetical syllogism was the early Peripatetics one. If it was the second, then the justification of the inference of A's not being from B's not being would have been derivative of that of the first hypothetical syllogism. But note that in either case there will be no hypothetical syllogism for which a sentence of the kind 'If A is not, B is not' is required as premiss, and that we can thus trust that Galen reported the ancients correctly, when at *IL* 3.1 he gave the three possibilities 'A ὄντος B εἶναι', 'A μὴ ὄντος B εἶναι' and 'A ὄντος B μὴ εἶναι' for hypothetical premisses, but not a fourth, 'A μὴ ὄντος B μὴ εἶναι'. It is only later, when the hypotheticals are conceived of as propositions rather than premisses, that the fourth possibility naturally suggests itself, from a combinatorical point of view; and indeed we find it added in Ammonius and Boethius.[54]

No question of reduction arises in the case of the dividing hypothetical syllogisms. Take

A or B (is/holds).	hypothetical premiss
Now A (is/holds).	probative premiss
Therefore B (is/does) not (hold).	conclusion

Here the hypothetical premiss is dividing (κατὰ διαίρεσιν), the division (διαίρεσις) being stated to hold between the things A and B. There is indicated to be a dependency between A and B, which can e.g. also be described as 'A ὄντος, B μὴ ἐστι' (Galen, *IL* 3.1, 3.5). Once A has been proved, this dependency allows the transition from A's being to B's not being. (The case of the other type of dividing hypothetical syllogism works *mutatis mutandis* in the same way. In this case, the dependency would perhaps rather be described as 'A οὐκ ὄντος, B ἐστι', to make apparent the possibility of a transition from A's not being to B's being.) The relation being one of exhaustive and exclusive disjunction, one could imagine that each phrase indicates one half of the – symmetrical – dependency.

We can then add some further characteristics of the hypothetical syllogistic of the ancient philosophers to the list of Section 1. It seems confirmed that for the ancients the ὑποθετικαὶ προτάσεις were conceived of as hypothetical premisses, i.e., as items whose only function was that of premisses in hypothetical syllogisms; furthermore, that in the hypothetical premiss of every hypothetical syllogism a relation of dependency is indicated (as stated at *IL* 3.1 and 3.3), which is based on either a relation of connection or a relation of division, and which is – canonically – expressed in the form of a conditional or a disjunction, respectively. And it is this relation of dependency (plus, in one case, possibly contraposition of argument) which

52 Cf. *IL* 8.2: κατὰ μέντοι τὰς ὑποθετικὰς προτάσεις οἱ μὲν ἄλλοι πάντες οἱ ῥηθέντες ἀρτίως ἀναπόδεικτοί εἰσι καὶ πρῶτοι πλὴν τοῦ προσλαμβάνοντος μὲν τὸ τοῦ λήγοντος ἀντικείμενον, ἐπιφέροντος δὲ τοῦ ἡγουμὲνου τὸ ἀντικείμενον· οὗτος γὰρ μόνος ἀποδείξεως δεῖται.

53 On the other hand, at *IL* 8.2 (see previous note) Galen uses mainly Stoic terminology, and this suggests that he is drawing from a source later than the early Peripatetics.

54 Amm. *Int*. 3.11-15 and Boethius *HS* 1.2.7; cf. Philop. *APr* 243.11-13.

justifies the transition from the second premiss to the conclusion, or the inference of the conclusion, and which gives the hypothetical premisses (and the hypothetical syllogisms) their name. As to the origin of this early Peripatetic hypothetical syllogistic, it may be part of Aristotle's pupils' attempt at systematizing their teacher's 'other syllogisms from a hypothesis', perhaps with recourse to some pertinent passages from the *Topics*.[55]

On the negative side, we can note that there is no mention ever of the question of whether the hypothetical premisses were thought to have truth-values, and if yes, what their truth-criteria were. Nor is there sufficient information to allow us to decide whether the things (πράγματα) that are relevant for the syllogistic form of the arguments were generic things like human and animal, which hold of other things, or rather something like states of affairs, which simply hold or obtain. In sum, there is no evidence that we have anything like a 'propositional logic' even in a very wide sense, that is where the units relevant for the logical form of the arguments are propositions or whole sentences.

4. A later elaboration of the hypothetical syllogistic of the ancients

I now add a further point of support for my claim that Galen's ancient philosophers were early Peripatetics including Theophrastus and Eudemus, and that they had the basic theory I have sketched. Boethius, in his treatise *On Hypothetical Syllogisms*, maintains that he elaborates on Theophrastus' and Eudemus' hypothetical syllogistic (*HS* 1.1.3-4); and his theory of those hypothetical syllogisms that are formed from one hypothetical and one categorical premiss (*HS* 2.1.7-4.3; 3.10.3-11.7) is indeed an elaboration of a theory that shares all the main characteristics of the one I have reconstructed for Galen's ancients.[56] Moreover, among later ancient authors Boethius (in *HS*) is – almost – unique in doing so.[57] Here are the main points of agreement:

1. By *propositio* (which is his translation of πρότασις) Boethius often intends premiss, not proposition.[58]

2. The theory contains only conditional and disjunctive hypothetical premisses; in particular, there is never a formulation of a dividing or disjunctive premiss as a negated conjunction.

3. The two types of premisses are referred to as *per conexionem propositio* (*HS* 3.11.6, cf. 1.3.4, 2.4.3) and as *per disiunctionem propositio* (*HS* 3.10.3, 4, 5). These two expressions could be translations of ὑποθετικὴ πρότασις κατὰ συνέχειαν and ὑποθετικὴ

55 I have pursued the question of the origin of Peripatetic hypothetical syllogistic in detail in my 'The development of *modus ponens* arguments in antiquity', *Phronesis* (2002, forthcoming).

56 The elaboration consists in the main in Boethius' working through all the permutations one obtains for the four types of syllogism by substituting negative (component) propositions for either or both of the affirmative ones of the 'basic forms'. I have a hunch that this elaboration goes back to Porphyry, but as I am unable to prove this, I just mention it here.

57 The only other passage I know of is a long scholium to Aristotle's *Analytics*, which is closely parallel to Boethius' *HS*; cf. my 'A Greek parallel to Boethius' *De Hypotheticis Syllogismis*', *Mnemosyne* IV.3, 285-300.

58 Cf. *HS* 2.1.1 *propositio vel sumptum* and *HS* 2.2.1, 2.2.2, 2.2.3, 2.10.6 and 2.11.1, where Boethius uses *propositio* for the first (and hypothetical) premiss, and *assumptio* for the second (and categorical) premiss of hypothetical syllogisms.

πρότασις κατὰ διαίρεσιν. (Boethius does not add 'hypothetical' (condicionalis), since in books two and three of *On Hypothetical Syllogisms* he talks about hypothetical premisses only.)

4. 'si A non est, B est' is (said to be equivalent to) a disjunction 'aut A est aut B est' (*HS* 3.10.4 end). This mirrors the relation between the description of the consequence in 'A μὴ ὄντος, B ἐστι' and the disjunctive formulation in the hypothetical premiss 'ἤτοι A ἢ B' in Galen.

5. Boethius introduces four basic types of hypothetical syllogisms. They all consist of one hypothetical and one categorical premiss. Two are composed with a connecting hypothetical premiss, two with a dividing one.[59]

6. Most of the examples Boethius uses are pre-Stoic (and Peripatetic, that is, they resemble those in Aristotle's *Prior Analytics* and *Categories*): e.g. *si homo est, animal est, aut aeger est aut sanus*.[60]

To sum up, the theory which Boethius announces as an elaboration of Eudemus' and Theophrastus' theories is closer to the position that Galen attributes to the ancients than any other extant source.

5. Solving two difficulties in Galen's *Institutio Logica*

If we assume that Galen's ancient philosophers were early Peripatetics, including Theophrastus and Eudemus, and that they had the minimal(ist) theory of hypothetical syllogistic I have argued they had, then we can shed some light on two difficulties in the *Institutio*, one at *IL* 3.3 and one at *IL* 3.5.

At *IL* 3.5 we learn that those philosophers who focus on the nature of the things rather than on linguistic form, i.e. the non-Stoic logicians, would call the sentence 'If it is not day, it is night' a disjunction, and not, as we may have expected, a conditional. The reason for this should now be apparent. The relation between things on which an inference with this sentence as premiss would be based is a division (διαίρεσις); and the formulation 'If it is not day, it is night' reveals the relation of dependency (in a division) which makes possible the transition (μετάβασις) from second premiss to conclusion, and which was described in the account of the dividing hypothetical premiss as 'τίνος οὐκ ὄντος, τὸ ἐστι' (*IL* 3.1).

At *IL* 3.3 we learn that the ancients called a premiss which we use when we think that since this is not, this is, a dividing (i.e. disjunctive) premiss; and thus not as we may have expected, a connecting (i.e. conditional) premiss. Again, the reason should now be clear. The relation at issue is a division (διαίρεσις), and the passage describes what we think in terms of the relation of dependency which enables us to make the transition from one thing's not being to another thing's being. It does not give us the canonical (or any) form of dividing hypothetical premisses. In fact, we do not know whether the early Peripatetics had something like canonical formulations for hypothetical premisses. If they did, they may well have

59 He distinguishes eight (or ten) subtypes for each of these four. There is no parallel to this sub-distinction in our sources for early Peripatetic hypothetical syllogistic.

60 Cf. *HS* 2.2.2 *si homo est, animal est*; 2.2.3 *si est nigrum, album non est*; 2.2.6: *si animal non est, non esse hominem*; 2.3.3 *si est homo, animal est*; 2.3.5 *si est nigrum album non est*; 3.10.6 *aut aeger est aut sanus*, and similarly 3.11.1-2, 4-5.

involved the use of 'if' for connective premisses and of 'or' for dividing premisses. In any case it seems that the Peripatetics did not have a syntactic definition of hypothetical premisses. What determined whether a premiss is connecting or dividing is the relation that is assumed to hold between the things: when this relation is a division, the premiss is dividing; when it is a connection, the premiss is connecting. In this sense, the approach of the early Peripatetics in Galen is semantic, not syntactic.

Department of Philosophy, Yale University

ON THE AUTHENTICITY OF AN 'EXCERPT' FROM GALEN'S *COMMENTARY ON THE TIMAEUS*

DIETHARD NICKEL

These remarks relate to the overall theme of 'The unknown Galen' only in a strictly negative sense, for it will emerge that the excerpts under consideration are unknown only insofar as they are not by Galen, and Galenic only insofar as they are already familiar. It is indeed true that none of the surviving fragments from Galen's *Commentary on Plato's Timaeus* were included in C. G. Kühn's collected edition of Galen's works. The earliest publication of any of them was in Charles Daremberg's monograph of 1848, and until now the definitive edition of these fragments has been that of Heinrich Otto Schröder, printed in 1934 as a *Supplement* to *CMG*.[1] In 1992 Carlos J. Larrain published what he claimed were new excerpts from this commentary, but whether there is any justification for regarding them as authentic is questionable.[2] If the answer were positive, it would indeed constitute some previously 'unknown Galen', which would not only add to the contents of Kühn's edition, but considerably extend the scope of the extant fragments from this commentary. But one must have reservations about this.

The text in question is preserved in the 14th-century manuscript Scorialensis Gr. Φ III 11, fols 123r-126v, to which Larrain was the first to draw attention.[3] Parallels with Galen's writings and with Plato's *Timaeus* led him to the conclusion that he was dealing with an excerpt from a work of Galen on this dialogue, specifically 'a summary of the first two books of the *Commentary on the Timaeus*'.[4] The fact that the excerptor twice identifies Galen as the source of his material confirmed Larrain in his attitude. He expressed his considered opinion clearly in the following terms: 'The mode of expression and the philosophical framework unmistakably correspond to those of Galen'.[5] The fact that on examination it appears that some of the statements in this text are much less common elsewhere in Galen simply makes Larrain even more convinced that they provide new insights into Galen's thinking.[6]

Even these concisely sketched propositions provide several points for critical consideration. That Galen is referred to twice, once by name and once via the title of a Galenic

1 C. Daremberg, ed. and trans., *Fragments du Commentaire de Galien sur le Timée de Platon* (Paris and Leipzig 1848); H. O. Schröder, ed., *Galeni In Platonis Timaeum commentarii fragmenta, CMG* Suppl. I (Leipzig and Berlin 1934).

2 C. J. Larrain, *Galens Kommentar zu Platons Timaios*, Beiträge zur Altertumskunde 29 (Stuttgart 1992); cf. the favourable review by I. Garofalo, *Gnomon* 67 (1995) 645-46.

3 For the date of this manuscript, see Larrain, *Galens Kommentar*, 226.

4 Ibid., 7-8.

5 Ibid., 8-9.

6 Ibid., 9.

tract,[7] need imply nothing more than that the author of the text is harking back to certain Galenic ideas, without saying anything definite about the precise context of his borrowing. To what extent thought and expression coincide 'unmistakably' (*eindeutig*) with what we find in indubitably authentic works of Galen will be subjected to examination in individual cases. Should any statements found in the text turn out to have no parallel in Galen, then we move on to the question of whether their meaning can be made to agree with Galen's position as revealed in other sources.

The constraints of space require that my investigation be restricted to a small selection of extracts from the text, which I propose to examine in more detail from the aspects I have listed. There are two passages which are broadly representative of the text as a whole, so that analysis of them allows us to draw more generalised conclusions. The decisive criterion for my choice of extracts was that they should deal with matters in which Galen demonstrably took a considerable interest. In this way, a comparison between the new material and already familiar passages is most likely to lead to a conclusion as to whether Larrain's thesis, that they are both written by the same author, is admissible.

The subject of Larrain's fragment 6 is 'the relationship between body and soul in the embryo'. Only by formulating the topic in such general terms can one establish any correspondence as regards content between this text and any known statements of Galen. The passage recalls the one in Plato's *Timaeus* 43 A, which is quoted verbatim, and which begins with the statement that 'they [the newly created gods] bound the circuits of the immortal soul in the ebbing and flowing tide of the body' (εἰς ἐπίρρυτον σῶμα καὶ ἀπόρρυτον; lines 6-7).[8] With regard to this quotation, which conveys only a vague idea of its subject matter, the author states [lines 1-5] that not even the Platonists themselves agreed on how and when 'the incorporeal soul achieved union with the body', whether it was at the moment of birth, at the very outset, i.e. at conception, or at some intermediate stage.[9] The general thrust of the question concerns the ensoulment of the embryo, in particular the exact circumstances and timing of this process, which is envisaged as the entry into the body of a soul originating outside it. This subject matter is familiar from the Neoplatonic treatise Πρὸς Γαῦρον περὶ τοῦ πῶς ἐμψυχοῦται τὰ ἔμβρυα wrongly associated with Galen's name.[10] Galen himself, however, regarded the problem of ensoulment as irrelevant.[11] As may be seen from the detailed discussion in his treatise *De foetuum formatione*, what exercised Galen was the question of whether the soul or one of its parts could be regarded as the power responsible for the development of the embryo.[12] If this were the case, then in his estimation the soul would have to have been in contact with the matter of the developing body right from the

7 Fragment 1,1: p. 21; fragment 28,4-5: p. 165 (ὡς ἐν τοῖς περὶ τῆς τῶν ἁπλῶν φαρμάκων δυνάμεως ἐπιδέδεικται).

8 Galen quotes these words at *Quod animi mores* 4: p., 42,11-17 Müller = IV 780,9-13 K., commenting that Plato is here thinking of the moisture of the substance of the new-born.

9 In Larrain's printed text, κυηθῆναι, line 3, should be corrected to <ἀπο>κυηθῆναι, cf. ἀποκυηθὲν, line 11.

10 K. Kalbfleisch, 'Die neuplatonische, fälschlich dem Galen zugeschriebene Schrift Πρὸς Γαῦρον περὶ τοῦ πῶς ἐμψυχοῦται τὰ ἔμβρυα', in *Anhang z. d. Abh. d. königl. Preuß. Akademie d. Wiss.*, Berlin 1895.

11 Galen mentions 'the so-called ensoulment' (τῆς καλουμένης ἐμψυχώσεως) only once, in passing, at *De propr. plac.* 15,2: *CMG* V 3,2, 118,8, with V. Nutton's note at p. 213.

12 Galen, *De foet. form.* 6,13-16. 25. 29-32: *CMG* V 3,3, 96,16-98,4; 100,24-29; 102,28-106,2.

outset. Galen acknowledges that he cannot in fact answer the question as to what is responsible for the embryo's development.[13] In this connection, he states that for him there is simply no doubt of 'the rational soul which exists after birth and throughout life' not being this power responsible for development.[14] But this does not mean, as Larrain interprets it, that this soul 'will enter us' only postnatally, but that after birth it demonstrably exists and gradually begins to exercise its functions.[15] As Galen puts it, 'children are clearly not yet fully developed at birth, whether from the physical point of view, or with regard to the highest, i.e. the rational part of the soul'.[16] This statement, however, does not imply that in Galen's view the rational soul only enters the body at this stage, but that he envisaged it as already present, although its powers still needed to develop. The problem of ensoulment which is raised in this 'excerpt' interested Galen not at all. It follows then that this does not constitute a point of contact between the dubious text and the authentic work of Galen. This conclusion does not, however, exclude the possibility that Galen may have been setting out the divergence of opinion among Platonists on this point.

The passage from Plato quoted in the 6th fragment describes the circuits of the soul as being 'bound in a mighty stream' (lines 7-8). This metaphor is then interpreted as an allusion to the fluidity of the substance providing the raw material for the body, 'which takes its initial form (τὴν ... πρώτην σύστασιν) from menstrual blood and semen' (lines 9-11). Thus far the fragment corresponds exactly with Galen's concepts and his style of expression.[17] But when we come to the point where the body is said 'after birth (ἀποκυηθὲν) to resemble cheese just on the point of solidifying' (τοῖς νεωστὶ πηγνυμένοις τυροῖς παραπλήσιον; lines 11-12) we meet a comparison which, while it does appear similarly expressed in genuine Galenic texts, is used there to refer either to a different stage of development, or to specific component parts of the body. Thus in *De semine* Galen compares the embryo in the first stage of its formation (τὸ πρῶτον κύημα) with 'milk which is just beginning to curdle' (παραπλήσιον οὖν ἐστιν ἄρτι πηγνυμένῳ γάλακτι) and which for that reason no cheesemaker would attempt to shape.[18] In *De temperamentis* it is the bones of a young slaughtered animal which are compared to 'cheese which is just starting to set' (τὸ δ' ὀστῶδες γένος ἅπαν ἄρτι πηγνυμένῳ τυρῷ ἐμφερές).[19] The similarity between the passages would seem in this case to indicate, not that they are from the same hand, but that the 'excerpt' constitutes a deliberate imitation.

The fragment concludes with the observation that the loss of moisture from the body through innate heat is compensated by the influx of some other substance (lines 12-16); 'For

13 Ibid. 6,31: *CMG* V 3,3, 104,15-16.

14 Ibid.: *CMG* V 3,3, 104,22.

15 Larrain, *Galens Kommentar,* 69, note 7.

16 Galen, *De foet. form.* 3,29: *CMG* V 3,3, 78,7-8.

17 See, e.g., *De sem.* II 1,63: *CMG* V 3,1, 158,26-27. The phrase ἡ πρώτη σύστασις (τοῦ ζῴου, τοῦ κυήματος, sc. τῶν ἀγγείων, τῶν ὑμένων) is also attested, e.g. I 9,1; II 5,70: *CMG* V 3,1, 92,19; 194,21-22; *De foet. form.* 2,9: *CMG* V 3,3, 62,20. For Galen's view of the function of seed and menstrual blood in the formation of the embryo, see D. Nickel, *Untersuchungen zur Embryologie Galens,* Schriften zur Geschichte und Kultur der Antike 27 (Berlin 1989) 29-40.

18 *De sem.* II 5,29-30: *CMG* V 3,1, 186,2-5.

19 *De temp.* II 2: p. 44,11-14 Helm. = I 578,17-579,3 K.

it was of course necessary that it (i.e. the body) possesses connate heat (θερμότης σύμφυτος) – for this is what, in his (i.e. Plato's) view, causes the inmixed moisture to evaporate (διαφορεῖσθαι) – and therefore it was necessary that the gods, who put the body together, should see to it that an influx (ἐπιρροή, cf. ἐπίρρυτον σῶμα, line 7) of some other substance replaces what is lost so that the body is not destroyed.' As far as content goes, this fits in with Galen's ideas.[20] But in a genuinely Galenic fragment from the *Commentary on the Timaeus* the same material is presented in the following terms:[21]

> It is, however, necessary that the body, being controlled by these [i.e. the more active elements, fire and pneuma or air] has not only many forms of evacuation but also [cutaneous] respiration and effluxes, processes which are not perceptible to the senses (τὰς ἀδήλους αἰσθήσει διαπνοάς τε καὶ ἀπορροάς); it follows then that there must be a process of nourishment, whereby what has evaporated (τὸ διαφορούμενον) from its substance is replaced, and for the purposes of this nourishment the gods have provided plants.

This discrepancy in the mode of presentation means that the most one can say with regard to this 'excerpt' is that its author borrowed ideas from the works of Galen.

Let us now turn to the 14th fragment.[22] Its theme is one that Galen covers in great depth in the course of his discussion of Aristotle, namely whether the heart should in fact be regarded as the origin of the nerves. The relevant passage follows in full:

> It shows that all the nerves of the animal have their origin in the brain, and from them small outgrowths also pass to the heart. For this too must have some share in the principle that is above (τῆς ἄνωθεν ἀρχῆς) because it has to perform some service for it, as the following will show. At the points where the membranes [i.e. the heart-valves] emerge, there are clearly fixed some fine, disordered (?)[23] ligaments (συνδέσμους ... τινας ... λεπτοὺς καὶ ἀνόμους): Aristotle, who observed them, thought they were nerves in the heart, not knowing that a nerve is extremely sensitive, whereas a ligament is extremely insensitive. The tendons (τένοντες) – these are the nerve-like (νευρώδεις) terminations of these muscles – have a composition midway between that of nerves and ligaments. If one is going to agree with him that these ligaments in the heart are in fact nerves, it is also clear that there is nothing leading from it[24] to any part of the body, as there is from the brain and spinal cord to all parts. So far from the origin of the nerves being in the heart, not a single one of the very tiny nerves that come to it can be seen dividing up into the whole (heart), as is visible in many other parts of the body, but as soon as they are implanted[25], they become immediately quite invisible, as one can see even in the largest of animals.

Here both subject matter and argumentation come over as authentically Galenic. An unknown Galen perhaps? On the contrary, an extremely well known Galen, carved up and

20 See Galen, *De foet. form.* 5,16: *CMG* V 3,3, 88,18-21, with my note *ad loc.*, p. 148.

21 Galen, *In Plat. Tim. comm.* III 2: *CMG* Suppl. I, 11,4-8.

22 Larrain, *Galens Kommentar*, 114.

23 Instead of Larrain's ἀνόμους, which gives no sense, one should write ἰνώδεις (fibre-like), cf. the phrase συνδέσμους λεπτοὺς ἰνώδεις at *Admin. anat.* V 4: p. 297,18 Garofalo = II 500,14 K.

24 αὐτῶν, line 13 (i.e. τῶν νεύρων) must be changed to αὐτῆς (sc. τῆς καρδίας).

25 Reading ἐμφῦναι for ἐκφῦναι, line 18; cf. Garofalo's review, above, note 2, 646.

reassembled. The Galenic texts which form the basis of this passage can be precisely identified. The excerptor has transferred extracts from Book I of *De placitis Hippocratis et Platonis* (chapters 7, 9 and 10) and from Book VI of *De usu partium* (chap. 18), virtually sentence for sentence, and phrase for phrase. The expositions in these source texts have been drastically shortened and simplified, the order of the statements shifted about, and occasionally the wording changed.

De placitis Hippocratis et Platonis is the source of the statements that the brain is the origin (ἀρχή) of all nerves, and that the heart too has a small share in them, although the fragment talks of 'small outgrowths' instead of 'a small (nerve)'.[26] The information about the existence of ligaments to the heart valves, which Aristotle is supposed to have erroneously interpreted as nerves, can also be traced back to *De placitis* (lines 5-7 = I 10,3-4.5: *CMG* V 4,1,2, 96,18-21. 22-24). The excerptor refers to these by the simplified term 'ligaments', whereas Galen had formulated his material more carefully, adopting from Herophilus the expression 'nerve-like strands' (νευρώδεις διαφύσεις), whose use corresponds to that of the ligaments. The remarks about the contrasting capacities for perception of nerves and ligaments are similarly drawn from *De placitis* (lines 8-9 = I 9,2: *CMG* V 4,1,2, 94,14-16). Here again the formulation has been modified slightly: according to the 'fragment' the nerve is 'extremely sensitive'(αἰσθητικώτατον), the ligament 'extremely insensitive' (ἀναισθητότατος); as Galen presents the contrast, the nerve 'provides perception' (αἴσθησιν ... παράγον), the ligament is 'insensitive' (ἀναίσθητος).

The next two sentences of this passage demonstrate particularly clearly its derivation from *De placitis*, because not only does the content of the remarks coincide, but the wording also largely corresponds. In the source the two sections of text are not directly contiguous. In the 'excerpt' we read (lines 9-11): 'The tendons (τένοντες) – these are the nerve-like terminations of the muscles (νευρώδεις τελευταὶ μυῶν) – have a composition midway between that of nerves and ligaments (ἐν τῷ μεταξὺ νεύρων τε καὶ συνδέσμων)'. In *De placitis* Galen expresses the same subject matter like this (I 9,2.7: *CMG* V 4,1,2, 94,16-17. 29-30): 'A tendon (τένων), finally, is the nerve-like end of a muscle (πέρας ... νευρῶδες μυὸς)... The tendons are made up of ligaments and the harder nerves, and they thereby occupy, understandably enough, a position midway between the two (ἐν τῷ μέσῳ ... ἀμφοῖν)'.

The next passage from this fragment is formulated in a manner exactly parallel to its source (lines 11-15): 'If one is going to agree (συγχωρήσει) with him (sc. Aristotle) that these ligaments in the heart are in fact nerves, it is also clear that there is nothing leading from it to any part of the body (πρὸς οὐδὲν μέρος τοῦ σώματος ἀπ' αὐτῆς φερόμενον), as there is from the brain and spinal cord (ὡς ἀπ' ἐγκεφάλου τε καὶ νωτιαίου) to all parts. Far from the origin of the nerves being in the heart (νεύρων ἀρχὴν κατὰ τὴν καρδίαν ὑπάρχειν) ...' In this case the resemblance to the corresponding passage in *De placitis* leaves not the slightest shadow of doubt about the text's source. There (I 10,6: *CMG* V 4,1,2, 96,24-28), the wording reads: 'But if one agrees (συγχωρήσειεν) that they are nerves, and not merely nerve-like structures, it still does not necessarily follow that the heart is the source of the nerves (τὴν καρδίαν ὑπάρχειν νεύρων ἀρχὴν), because we cannot point to any nerve which leads from it (i.e. the heart) to any part of the body (εἰς μηδὲν μόριον τοῦ

26 Lines 1-2 = *De plac. Hipp. et Plat.* I 7,55: *CMG* V 4,1,2,90,24-25; lines 2-3 = I 10,1: *CMG* V 4,1,2, 96,12-14.

σώματος ἀπ' αὐτῆς φερόμενον), in the way we have already demonstrated happens in the case of the brain and the spinal cord (ὡς ... ἀπ' ἐγκεφάλου τε καὶ νωτιαίου).'

The excerptor was even more cavalier in his handling of the second source for fragment 14. It is beyond all dispute that *De usu partium* is the text he was using here, but he deals with some of its statements in a distinctly arbitrary manner. The assertion which appears in the 'excerpt' (lines 15-17) to the effect that one cannot see any of the nerves 'dividing up into the whole (heart)' (σχιζόμενον εἰς ἅπασαν αὐτήν), does indeed have a parallel in the source text[27] (*De usu partium* VI 18: I 364,6-7 Helmr. = III 500,5-6 K.: νεῦρον δ' οὐδὲν φαίνεται κατεσχισμένον εἰς αὐτήν, sc. τὴν καρδίαν); but in Galen it is followed by examples of other organs where this is also the case, such as the liver, the kidneys and the spleen (364,7-8 Helmr. = 500,6-7 K.). The excerpt, however, turns the parallel into a contrast, 'such as may be seen in the case of numerous other parts of the body' (lines 17-18). A similar reversal awaits us in the following sentence: 'even in the largest of animals' (καὶ ἐπὶ τῶν μεγίστων ... ζῴων), the excerpt informs us, these tiny nerves are no longer detectable once they have become implanted into the heart (lines 18-19); in contrast, Galen had stated that 'at any rate in the larger animals' (ἐπὶ γοῦν τῶν μειζόνων ζῴων) perceptible offshoots of these nerves may be observed, which are implanted into the heart itself (364,10-12 Helmr. = 500,9-11 K).

Consideration of these selected extracts has produced a variety of findings: discussion of a theme with which Galen demonstrably did not concern himself, dependence on Galen's ideas, and, most striking of all, the compilation of extracts from his extant works. The relevant passages of the authentic writings of Galen were familiar to Larrain, but he failed to draw the correct conclusion from his knowledge.[28] Instead of identifying them as the actual sources of his excerpts, he assessed them only in general terms as evidence of Galenic ways of thinking, into which the supposed extracts from the lost *Commentary on the Timaeus* could be more or less easily fitted under the guise of uniformity. From the textual analysis I have just undertaken, it will no longer be possible seriously to maintain that we may be dealing here with fragments of this commentary. It must be acknowledged that not all the passages provide equally good starting points for a refutation of Larrain's thesis, especially when they cover topics which have no equivalents in Galen's established works for comparison. But in these cases the onus is on the editor to demonstrate their authenticity, in the sense of their being fragments from Galen's *Commentary on the Timaeus*. He has, in fact, not even attempted to undertake such a proof. The value of these texts for the study of Galen is very small. In those cases where they can be shown to convey Galenic material we do not need them, because the genuine writings on which they are based are available as more reliable sources, and what they contain in terms of unfamiliar material is of dubious authenticity. At best, they may be of some evidential value in the context of *Wirkungsgeschichte*.

Berlin-Brandenburg Academy of Sciences

27 G. Helmreich, ed., *Galeni De usu partium*, 2 vols (Leipzig 1907-1909).

28 Larrain, *Galens Kommentar*, 9, acknowledges that there are many correspondences in this 'excerpt' with *De plac. Hipp. et Plat.*

GALEN *ON THE UNCLEAR MOVEMENTS*

ARMELLE DEBRU

Throughout his life Galen was an anatomist. He was so already as a student, and his first stay in Rome saw his splendid anatomical demonstrations and the production of his first important books in this field. His interest did not weaken during his second stay, while he was writing or finishing such major works as *On the use of the parts* and *Anatomical procedures*, as well as many other anatomo-physiological treatises. Until extreme old age he worked for the transmission of this knowledge to experienced colleagues, to beginners, and to those with a non-professional interest. He rewrote some books of *Anatomical procedures* after the fire of 192, and he made a great use of anatomical considerations in his discussions about the soul in his final book *On my own opinions*. The word 'anatomy' covers not only his research into bodily structures but also his explanation of their use, which has inseparable medical and philosophical implications. Galen was, at one and the same time, an anatomist and a philosopher; to coin a word, he was an anatomo-philosopher.

A new anatomical work has recently been (re-)introduced into the Galenic Corpus, *Περὶ τῶν ἀπορῶν δυναμέων*, which was known in Latin under a variety of titles, including: *De motibus obscuris, De motibus manifestis et obscuris, De motibus dubiis* and *De motibus liquidis.*[1] This treatise was known and used by a Christian bishop in the late 4th century, Nemesius of Emesa, in his treatise of Christian anthropology *De natura hominis*. Nemesius adopts Galen's categories and thesis, for example, the distinction between voluntary and involuntary motion and faculties. But for him, however, God did not completely share in both domains, but 'in his supreme forethought, He has woven all that proceeds from the soul together with whatever springs from nature, and the latter with the former'.[2] This statement is immediately followed by an example – defecation – given in *De motibus obscuris*, and there are other allusions to this treatise. Later, it was translated into Syriac and into Arabic by Hunain ibn Ishaq, and it interested Arabic commentators and philosophers, such as Avicenna and Averroes, in a variety of ways.[3] But until now, *De motibus obscuris* was

1 A single fragment has been found in Greek, in a manuscript of byzantine theology, and edited by C. J. Larrain, 'Ein bislang unbekanntes griechisches Fragment der Galen zugeschriebenen Schrift *Περὶ ἀπορῶν δυναμέων* (*De motibus dubiis*)', *Philologus* 137 (1993) 265-73.

2 Nemesius, *De natura hominis* 252, 5: tr. W. Telfer, 373. For further comparisons, see C. J. Larrain, 'Kommentar zu Galen *De motibus dubiis* in der mittelalterlichen lateinischen Übersetzung des Niccolò da Reggio', *Traditio* 51 (1996) 1-41 (9-11).

3 For these translations, see Larrain, 'Fragment', above, note 1, 265-66. An edition of the Arabic is being prepared by Gerrit Bos, see below, p. 143.

essentially known to scholars only in the arabo-latin translation of Marcus Toledanus (12th century), which was available in the collected editions of Galen in Latin from 1490 onwards and was included among the *Spuria* in Chartier's bilingual edition of the *Opera Omnia.* Carlos Larrain has recently edited for the first time the Greek-Latin translation of Niccolò da Reggio (14th century), and has published it with the translation of Marcus Toledanus.[4] Although both Latin translations are difficult to interpret, they make this treatise accessible again for wider discussion. It displays a remarkable anatomical knowledge, a mastery of discussion, its method, and a mixture of respect and disdain for the frontiers between medicine, morals and philosophy which characterises Galen's later work.

Galen's exposition of an anatomical problem whose central concept is that of *kinesis* (movement) is essentially a treatise on *aporia. Kinesis* is understood here in its physiological meaning, not in the larger Aristotelian one encompassing all kind of changes and activities, a meaning that Galen knows as well. On the other hand, *kinesis* includes activities classified as 'natural' when they are discussed in the treatise *On the natural faculties*, such as attraction, expulsion, retention. But all the movements considered in the *De motibus obscuris* are visible, evident (*liquidi*), implying some form of spatial movement. If they are said to be aporetic, doubtful or unclear, it is not because they are difficult to distinguish, for instance through being too small, but because they are difficult or even impossible to explain. In particular, they appear to contradict a rule which was accepted, so Galen claims, by all anatomists of his time as heirs to the Alexandrian (or at least Erasistratean) concept of movement. They believed that voluntary movement was exclusively accomplished by the muscles, which received nerves from the brain and the spinal cord. Contrary to this rule, the examples brought together in this treatise concern voluntary movements which are not accomplished by muscles. As these movements exist, they have to be explained. The question is whether they are voluntary or natural, whether they are accomplished by muscles or by another kind of structure. It is not easy to decide. As J. Mansfeld noticed when sharply criticising Galen on this point, the dominant physiological paradigm is that of voluntary acts accomplished by the muscular system; that of involuntary acts, which Galen wants to be completely separated from the preceding, lacks any physiological explanation.[5] According to Mansfeld, there can be no solution but only an hypothesis. This explains why *De motibus obscuris* talks a great deal about philosophical developments, which are not simply brought together but deliberately inserted into Galen's anatomical reflections.

The treatise begins with an awkward sentence; 'Anatomists call movements unclear those in which, although we see perfectly clearly that there is a movement in animals, we are entirely ignorant of the part by which it is accomplished (*a qua parte fit*) or, if we know that, we do not know how it is produced (*quomodo fit*).'[6] This is the *aporia*, although it would be better to speak about *aporiai*, difficulties, because there are several sorts of difficulty.

4 Id. C. J. Larrain, 'Galen, *De motibus dubiis*, die lateinische Übersetzung des Niccolò da Reggio', *Traditio* 49 (1994) 171-233. My references are to the pages in this edition.

5 J. Mansfeld, 'The Idea of Will in Chrysippus, Posidonius, and Galen', *Proceedings of the Boston area Colloquium in Ancient Philosophy* 7 (1991) 107-57.

6 MT = Marcus Toledanus, *De motibus dubiis* 191,5-12 Larrain, in Niccolò da Reggio's version (= NR): 'dubios motus vocant qui vacant circa anathomiam in quibus quamquam quidem (quod ? VN) est in animalibus motus evidenter videmus sed vel omnino nescimus vel a qua parte fit vel hoc scientes ignoramus quomodo fit'. The version of Marcus Toledanus (= MT) is unhelpful here.

Our ignorance may be provisional. Galen gives us several examples of temporary ignorance, and of findings which resolved his own and others' medical *aporiai* concerning the nature of movement. The need to define and distinguish between active and passive movement on a fundamentally physiological basis seems to have exercised many anatomists, and Galen begins by offering his own definitions.[7] The debate was particularly contentious over respiration, as Galen explains at length.[8] He himself had made important discoveries of considerable relevance to this question: several muscles, which were unknown to his predecessors, findings about respiration and the voice, the muscles of the eyelids, the recurrent laryngeal nerve, the muscles of the larynx etc., all discoveries that we find in other Galenic works.[9] In these no longer ambiguous cases, anatomy has produced results because it has found the organs, the ways, and then the nature of movements. *Aporia,* in this case, has an heuristic value. This is *positive* anatomy.

By contrast, *aporia* appears resistant to a resolution in two instances: when a voluntary movement as evident as the preceding ones does not seem to involve a muscle and when an involuntary movement takes place with the help of a muscle – two situations which are theoretically impossible but whose existence is attested by observation. It is easy to see that this *aporia* is based on a dichotomy: muscle versus non-muscle, voluntary versus involuntary, although in these dichotomies, which are strongly correlated, the second term is, as often, not the exact contrary of the first. But this time it helps if one can prove that the moving part is not a muscle, or that the movement does not belong to the class of voluntary actions. The purpose of the anatomical investigation is here primarily *negative.*

The core of *De motibus* is a catalogue of the doubtful instances when a voluntary or at least a part of a voluntary movement is not effected by a muscle. Galen's first examples are putting one's tongue out and the erection of the penis – both movements where muscles intervene but which also imply an increase of volume that no muscle can explain. Thus, there is an anatomical enigma:

> Cum ergo anathomiam inspicientes in hoc conveniant quod nullus motus obedit voluntati nisi motus lacerti tantum, nec aliquis lacertus lingue continuatus distendit eam donec adducat eam extra os, tunc merito in titubatione incidimus quando scire volumus qualiter actio fiat lingue.[10]

The problem would be solved, 'if a movement could be found which would follow the will and – at the same time – be executed without any intervention of a muscle'.[11] This leads Galen to state, with great hesitation, the hypothesis that a voluntary movement can take place

7 NR 191: 'Prima igitur inquisitio in motibus qui evidenter apparent est utrum sint actiones musculorum aut fiant secundum passionem' (NR). This question is debated in many other Galenic treatises, even in the *De moribus* (tr. Mattock 237).

8 NR Ch.1, 17-97: 196-197.

9 There is a simplified but interesting temporal scheme of *inventio* (NR 198): 'Mihi ... dubia videbantur antea ... Ita autem ... de motu superioris palpebre ignorare confessus sum; tempore autem post quomodo fit ... scripsi; eodem modo quomodo perit vox ... non reperi ... Tempore vero post inveni ... Dubium etiam erat propter qui ... inventum enim fuit a me ...'.

10 NT ch. 4: 210,77-211,84.

11 'Si motus aliquis posset in corporibus animalis inveniri qui voluntate fieret absque aliquo lacerto non incideret in hanc questionem aliquid ambiguitatis', MT 211,84-88.

without the instrument of will. He admits that this possibility could be solved by invoking imagination. What Galen thinks both organs do naturally is to attract pneuma, which suddenly arrives in their tissues: *impletur corpus rarificatum ex vento velociter ac leviter*, for instance, in the case of the penis, under the influence of images produced by imagination *(phantasiai)*.[12] This *repletio spiritus* could explain why some parts of the body are moved by other causes than the merely muscular, although this remains anatomically unproven.

After having shaped this methodological instrument, Galen examines a series of other aporetical movements. They are not characterised by their delicacy. But, as with the spiders, the lice or the crabs for Aristotle, all parts and actions of the body are for the anatomist worth inquiry and admiration.

First comes the question of voluntary vomiting and of the cow's rumination, which is also voluntary. This is mysterious because none of the organs involved in these phenomena, the oesophagus as well as the stomach, is composed of muscles, whereas the structure of their coats and the nature of their fibres permit only involuntary movements as shown in the treatise *On the natural faculties,* to which Galen refers many times. Then there come under scrutiny urination and defecation, as well as physical and emotional phenomena which transmit themselves from person to person for an unknown reason, like coughing, sneezing and laughing. Finally Galen devotes a long chapter (10) to a discussion of swallowing. Each of these movements poses a problem of explanation, for which Galen suggests his own solution.

In the case of voluntary vomiting, for example, the strict alternative 'either .. or' is prudently replaced by a more subtle solution, which talks of 'not only.. but also' (*non solum....sed etiam*), or of 'sometimes (voluntary) and sometimes (natural)'.[13] The possibility is also introduced that the voluntary act may somewhat help the natural movement in order to create mixed movements (*mixtas.... motiones*).[14]

Defecation allows Galen the opportunity to build another interesting model of the interaction of both types of faculties: while a natural expulsive movement takes place in the intestines, voluntary action consists in stopping the contraction of the sphincter. This question had apparently raised a considerable discussion.[15] Another question arising out of the stopping of action as voluntary movement seems to have fascinated logicians: can a cause stay a cause of something even when it stops acting?[16] The anatomical model of defecation has also other moral and juridical implications, as shown by two short delicious sketches, one about Apollo's reprimand against someone who did not visit an ill friend.[17] The other invites reflection on why should we condemn a deserter who did not take part in a battle?[18] Finally Galen adopts a

12 MT 12-13; 211 Larrain.

13 For example: 'Hoc igitur considerare convenit ne forsam et aliquibus aliis particulis insunt motus obedientes non solum impetibus animalium *sed etiam* fantasiis que secundum cogitationes fiunt' (NR 211,6-11*)*; about rumination: 'quoniam autem *non solum* musculis impetibus animalium inest moveri, manifeste est videre in animalibus ruminantibus' (NR 211,16-19).

14 NR 221,9-10. This idea, found in Nemesius, was thus taken over from Galen.

15 Here Galen alludes to other interesting anatomical debates – for example: 'Propter quod coniectaverunt quidam ... Querentes autem hii quid utique sint, discordati sunt ab invicem non parva discordia, alii quidem dicentes ... alii vero ... alii vero' (NR, 205,141-54).

16 NR 216,38-217,73.

17 NR 216,25-36; MT 216,28-37. The text of both versions is far from clear.

18 NR 217,56-73: MT 217,55-75.

graduated formula using two terms: *prima ratio* or *intentio* and *secunda ratio* or *intentio*. A movement may be voluntary by *prima intentio*, and involuntary by *secunda intentio*.

The other types of movement are thus covered by this kind of combinatory system of explanation. Among them, the distant transmission of an eye infection from one person to another, the appalling effect of seeing a snake, the transmission of envy or desire, examples that we find also at the beginning of *De moribus*. Only one movement remains without explanation or even hypothesis: the contagion of helpless laughter leaves Galen at loss.[19]

All these models of combinations of the two great instances of movement, even if they imply philosophical aspects as we will see, are a matter for anatomical explanation: we must know *qua parte fit* et *quomodo fit*. But there are also examples of what we could call negative anatomy. Anatomical knowledge, for instance, shows that the movement of the tongue or of the oesophagus does not come from the muscles, either because they do not reach the part involved, or because the part that is moved is composed of tissues whose coats and fibres do not go in the right direction. Most of these anatomical details figure also in the treatise *On the natural faculties*. But, as has been been pointed out by Mario Vegetti, what is at stake here is not teleology.[20] Anatomical observation is not limited to this negative result, but it also helps to explain how the natural movement, even in voluntary action, has other anatomical possibilities. For instance, it is because of its porosity and of the presence of large arteries in it that the tongue can receive a flood of *pneuma* and dilate for this reason. The visible movements of desire, appetite, and attraction can almost all be given a plausible and tentative anatomical explanation even if that explanation still remains problematic or unproven.

But there is a third level of movement, the most difficult and definitely aporetic, where there is a function without any apparent anatomical basis. We have already noticed that philosophical reflections were inserted at two places in the treatise: in the middle of the discussion about the tongue and the penis,[21] and later, in a passage about pain. We find the example of a philosopher suffering from his leg more during the night than during the day, a problem more appropriate to the genre of *Problemata* (Why do we suffer more during the night than during the day?).[22] The philosophical considerations developed here – a kind of theology of natural faculties – recall the treatises *On the natural faculties* and *On the formation of the foetus* on the one hand, and, on the other, chapters 10 and 11 in *On my own opinions*.

When he discusses the anatomical enigmas of the tongue and the penis, Galen tackles again the problem of the presence in the depths of the body of something (soul, Nature, providence

19 NR 223, 40-52: 'propter quid ... ridiculosum aliquod videntibus aut audientibus infert motum omnifarie, in[dis]solubile apparet'. Most of these examples also figure in Hippocrates, *De humoribus*.

20 M. Vegetti, 'Historiographical strategies in Galen's physiology', in *Ancient Historiographies of Medicine. Essays in medical doxography and historiography in Classical Antiquity*, ed. Philip van der Eijk (Leiden 1999) 383-95. For Vegetti (383, 388), the content of the *On the natural faculties* was for Galen 'a task that was certainly more difficult than the one he would have to face later when he was writing the more comprehensive work *On the Usefulness of the Parts* ... the problems Galen had to face in *On the Natural faculties* were complex, important and even more difficult if we consider that in this field he could not rely on the great tradition ranging from Aristotle's *De partibus* to the Alexandrian anatomy and to his own anatomical procedures. The points of reference were more uncertain. Besides, Galen is conscious of the precariousness of the teleological axiom he has introduced and defended so strongly.'

21 This insertion is substantial, NR 201, 33-206, 180.

22 NR 220,174-221,217: 'ex genere autem questionum est etiam propter quid eis qui occupantur circa alia flegmonantes particule minus dolent'.

and so on) and/or someone (the *creator*, to use the Latin term, although Galen often says God) who imparts knowledge to the parts of the body. First of all, each part knows what it needs for itself: *Hos scilicet providentiam habere de se ipsam unamquamque particularum, non solum est suasibile sed etiam verissimum dicere.*[23] This faculty is not at all altruistic; on the contrary it takes care of its own *cura*, it is selfish *(unumquodque membrum suiipsius intendit commodo).* Second, this knowledge is a kind of immanent cognitive faculty. It is not just tendency, desire, impulse, but the exact anticipation and adaptation of what is intended. At the very end of the treatise Galen repeats the Aristotelian example of the fish called 'cinodonta' whose stomach, as soon as the animal sees a prey, goes out of the mouth.[24]

But the paradigmatic example that we find many times in Galen and which was widespread at this time is the way children pronounce words without having any knowledge of the muscles of the tongue and their use.[25] The hypothesis expressed in this treatise is that he who formed our tongue, whoever he is, either remains in the parts he has formed, or has created these parts as animals, as Aristotle once expressed this idea.

The existence of an immanent cognitive faculty is a recurrent theme of this treatise. Envy or other emotions transmit themselves 'if members had a spirit'. At this point, Galen raises the same question as in *On the formation of the foetus: quis autem sit et quomodo inhabitat in corpore animalis?* At several moments, the cognitive faculty seems to be shared between both kingdoms: transmitted by the *hegemonikon* or lying in the depths of the body. The question of how this knowledge came here, whether from the *hegemonikon* or implanted in the parts themselves, and the extent to which it was autonomous, was intensively discussed among physicians and philosophers.

This leads Galen on to another problem, about the soul. We know the limitations and definitions imposed by Galen when he considered the substance and nature of the soul. But still he cannot avoid wondering whether it is the same soul as the *plasmatic* one which formed the embryo, which stays in the body (*sicut si esset presens creator qui plasmavit nos*) or if it is another one.[26] Here as elsewhere, it appears that things could be easier if it was the same soul that controlled knowledge, decision, and action. And if we read *De moribus,* we find that the first soul could easily absorb the second one or *vice versa.*

The question of the remaining faculty recurs also in *On my own opinions,* but here the question is slightly different: are the diaplasmatic faculty and the one which dwells in the body one and the same? Galen thinks that there is only one*: magis adhuc intelligo unam oportere esse animam quae et nos formet et quae singulis particulis nunc utatur*, but the question is really complicated. Galen resists the unification of the souls, saying explicitly that we have two souls, even if the one lying in the body, which is the servant *(que nostris*

23 NR 208, 237-40.

24 NR 233,394-401. See Aristotle, *HA* 591 b, already mentioned by Galen at 2,173-74 K.

25 Et enim et valde parvi infantes quando flectere aliquam partem aut extendere aut ad obliqua ducere volunt, confestim hoc agunt per proprios videlicet musculos illius motus ignotos non solum illis sed etiam pluribus medicorum. Sicut si esset presens creator qui plasmavit nos et scivisset singulum musculorum cuius gratia fecit (NR 200,23-201,35). The same example is found at IV 696-97 K.

26 NR 202,52-203,71. The same problem (re)appears in *On the formation of the foetus* and *On my own opinions.*

impetibus famulatur) of the will, gains dignity and perhaps even more attention than the *hegemonikon*.[27]

Because many other Galenic works are alluded to, this treatise gives sometimes the impression of being a cento, produced in the Christian centuries. Nevertheless, in spite of its considerable obscurities and its strange structure, the content may be said to be almost entirely Galenic.[28] As he so often did when returning in his late works to the problems of the soul already encountered in *On the natural faculties*, Galen here manifests his need to recapitulate his intellectual career. So we find here a history of his own discoveries, quotations from his previous works, and even some critical reflections on his theories. All these characteristics also indicate the importance of this lost and rediscovered treatise, which was several time quoted by Galen himself. It is of particular interest to the historian of physiology, for it confirms the centrality of the question of the movement as one that linked many anatomical, physiological, and philosophical problems about the body and the soul(s) within it.[29]

Université René Descartes, Paris

27 NR 203,64-68.

28 For example, after raising difficult questions about the soul, Galen devotes the last chapter to swallowing, not to aporetical philosophy.

29 I am grateful to Ivan Garofalo for drawing my attention to this treatise and for providing me with a copy of his own Italian translation, as well as to Vivian Nutton for improving my English.

THE BRAIN BEYOND KÜHN: REFLECTIONS ON *ANATOMICAL PROCEDURES*, BOOK IX*

JULIUS ROCCA

Reliquium noni una cum decimo, undecimo, duodecimo, tertio decimo, quarto decimo, et quinto decimo, etiam in Graecis exemplaribus desideratur.

Giuntine *Galen*, 1556

I

The second volume of Carl Gottlob Kühn's flawed but still indispensable edition of Galen contains only the first nine and a half books of his fifteen book anatomical masterpiece, *Anatomical procedures* (*De anatomicis administrationibus*).[1] The Greek text breaks off at book nine, chapter six, leaving approximately half of Galen's work in Greek.[2] For the remainder, we are in debt to the labours of the renowned Nestorian physician and translator, Hunayn Ibn Ishaq (d. 873). Hunayn renders Galen's often highly technical language with great skill – a daunting task made all the more so by the fact that it is highly unlikely Hunayn could have availed himself of the use of dissection to settle a disputed point.[3] But although an Arabic manuscript of the complete work was available in Western Europe from at least the early seventeen century, it was not until 1906 that Max Simon produced the first edition and German translation, and not until Duckworth's 1948 Linacre Lecture that the full extent of Galen's anatomical endeavours was revealed to the monoglot English physician.[4] In 1962, an English translation followed, based on Simon.[5] This made the whole work available in

* I am most grateful to the thoughtful criticisms and suggestions that were made by Armelle Debru and Heinrich von Staden. Special thanks are due to Vivian Nutton, whose assistance greatly improved this paper.

1 II 215-731 K. The Greek of Book IX occupies 707-31. Ἀνατομικαί ἐγχειρήσεις, 'anatomical undertakings', literally means work performed by hand. 'Practical Anatomy' is the apt translation given in *LSJ*, 475 s.v. ἐγχείρησις.

2 For other fragments in Greek, see R. K. French and G. E. R. Lloyd, 'Greek Fragments of the Lost Books of Galen's Anatomical Procedures' *AGM* 62 (1978) 235-49; and in Garofalo's *editio maior*, below, note 8, 582-91.

3 Hunayn's depiction of part of the terminations of the anterior ventricles does not quite fit in all respects when compared to the way in which Galen has oriented the description of the dissection of the brain which is given in the Greek part of Book IX.

4 M. Simon, *Sieben Bücher Anatomie des Galen*, two vols. (Leipzig 1906). W. L. H. Duckworth, *Some Notes on Galen's Anatomy*, Linacre Lecture, 6 May, 1948 (Cambridge 1949). It is debatable whether many physicians up to then had made use of Simon's 1906 translation. The complicated story of the fortunes of the Arabic manuscripts and of the attempts to publish an edition in the nineteenth century is chronicled by D. Gourevitch, 'Un livre fantôme: le Galien arabe de Greenhill', in *Les Voies de la Science grecque*, ed. D. Jacquart (Geneva 1997) 419-72.

5 W. L. H. Duckworth, M. C. Lyons, and B. Towers, *Galen. On Anatomical Procedures, The Later Books*, Cambridge, 1962. Although this edition is largely an English translation of Simon's German, Duckworth had the benefit of an Arabist (Lyons), and an anatomist (Towers), both of whom revised Duckworth's original translation. In this respect,

English, for in 1956 Charles Singer had provided a somewhat eclectic and erratic translation of the Greek text.[6] An Italian translation of the entire fifteen books was published by Ivan Garofalo in 1991, with a facing text of the Greek.[7] The same scholar completed his *editio maior* of both the surviving Greek and the corresponding Arabic in 2000.[8]

In its entirety, Book IX is a treasure house of Galen's anatomical science of the brain and nerves, and represents the culmination of Greek endeavour in this field. Not until Vesalius would anyone else approach its mastery.[9] In Book IX, Galen provides a detailed exposition of the anatomy of the brain and his experiments upon it, and outlines his work on elucidating the function of the nerves from the spinal cord. This paper highlights two particular topics, key elements of which are to be found principally in Hunayn's version: Galen's selective employment of animals for teaching purposes and his use of vivisection to demonstrate the hegemonic status of the brain.

II

Understandably enough, Galen's anatomical works have been more often cited than read for what they are. They are a mine from which several pertinent details about Galen's career may be extracted.[10] Such information is without doubt of great interest, for it provides details of Galen's own working background in Rome, even if we are dependent on his word for them. Yet this material is incidental in a treatise which is first and foremost a manual of dissection. Next to his therapeutic writings, it is the most practical text Galen ever wrote. In Book IX, chapter 11, Galen puts this into perspective by stating that it is intended as a successor to his *magnum opus* of teleological physiology, *On the usefulness of parts (De usu partium)*.[11] The Greek text of Book IX takes reader and student through the membranous coverings of the brain, its superficial and deep blood vessels, (especially the *torcular Herophili*) and those deeper structures, such as the *corpus callosum* and *fornix*, which, according to Galen support the underlying ventricles. It ends with an account of the fourth ventricle and its floor, the famous *calamus scriptorius* of Herophilus. In the second half of Book IX, the account of ventricular anatomy is completed, and the stage is set for Galen to demonstrate the nature of his experiments on the ventricles in living animals in order to determine their function. Whilst Galen does not claim originality for his discovery of the ventricles, he brings a special rigour

it is in some ways an improvement on Simon's work. Notwithstanding Garofalo's translation (below, note 7), a *CMG* edition of the entire fifteen books remains a desideratum.

6 C. Singer, *Galen. On Anatomical Procedures* (Oxford 1956). Singer's translation is not without its merits (although his introduction and notes are disappointing), but he took it upon himself to abbreviate the Greek text where, in his view, Galen had needlessly repeated himself.

7 I. Garofalo, *Galeno. Procedimenti Anatomici*, 3 vols (Milan 1991).

8 I. Garofalo, *Galenus. Anatomicarum Administrationum Libri qui supersunt novem. Earundem interpretatio arabica Hunaino Isaaci filio ascripta*, 2 vols (Naples 1986, 2000).

9 All citations from the second half of Book IX are taken from Duckworth's edition, supplemented, where necessary, by the texts of Simon and Garofalo.

10 For example, in the first chapter of Book I, Galen discusses how the text came to be written and subsequently revised and expanded at the urging of his friend, the ex-Consul Flavius Boethus. He is described as possessed of an 'ardent love of anatomical observation' (II 216 K.).

11 Galen's advice is that one must read the anatomical works in conjunction with his physiological texts (*De ord. lib. suorum*, XIX 54-55 K.; see below, note 13).

to their anatomy, elaborating an anatomical and physiological portrait of the ventricles which goes well beyond that first established by Herophilus and Erasistratus.[12]

Although Galen's *On vivisection* (*De anatomia vivorum*), is not extant, there is sufficient evidence from other works to determine which animals he employed, and for what purpose.[13] Galen used a great variety of animals for dissection and vivisection, especially primates, goats and pigs.[14] Many of Galen's dissections were carried out on primates; specifically, according to him, on five types of 'ape'.[15] Apart from πίϑηκος (the Barbary ape), the list includes λύγξ (an unknown tailed ape);[16] σάτυρος (perhaps *Macaca mulatta*, the Rhesus monkey); κυνοκέφαλος (dog-headed baboon);[17] and κῆβος (sometimes used as a synonym for σάτυρος).[18] The one most commonly used by Galen (and known to Aristotle) was the Barbary ape of North Africa (*Simia sylvanus* or *Macacus inuus*).[19] Galen made use of the ape in a way not dissimilar to how anatomy is usually taught today – that is, by prosected specimens, which can also be examined over a few days (allowing for varying periods of decomposition depending on the season and the tissues involved). The five types of apes formed for Galen part of a group of six classes of animals which he held were 'not far

12 What is extant does not allow one to conclude that the direction of their work necessarily parallelled that of Galen in all respects.

13 Cf. *De ord. lib. suorum* XIX 55 K. and above, note 11, referring to this work and another text on *Dissection* (*De anatomia mortuorum*), both now lost in Greek. The Giuntine *De anatomia vivorum* may, according to Ivan Garofalo, comprise genuine elements (G. Strohmaier, personal communication). See also I. Ormos, 'Bemerkungen zur editorischen Bearbeitung der Galenschrift 'Über die Sektion toter Lebewesen', in *Galen und das hellenistische Erbe*, ed. J. Kollesch, D. Nickel, AGM Beiheft 32 (Stuttgart 1993) 165-72.

14 Galen seems to have leant towards the Stoic view that animals were non-rational beings. On this reading, Galen maintained that animals suffer less than humans (cf. II 631-32 K.). On the treatment of animals in antiquity see J. Passmore, 'The Treatment of Animals', *J. Hist. Ideas* 36 (1975) 195-218, (198, 206-07). See also A-H. Maehle and U. Tröhler, 'Animal Experimentation from Antiquity to the End of the Nineteenth Century: Attitudes and Arguments', in *Vivisection in Historical Perspective,* ed. N. A. Rupke (London 1987) 14-47, although their notion (16) that Galen also avoided apes for vivisection in order to escape a charge of human vivisection is weak.

15 I employ Galen's generic term for primates, πίϑηκος, or 'ape', since he did not differentiate between what is today distinguished as ape and monkey. Cf. W. C. O. Hill, *Primates. Comparative Anatomy and Taxonomy*, 4 vols (Edinburgh 1953-74): vol. 1 (1953) 3-4; vol. 2 (1966) 2-10, 211-12; vol. 3 (1970) 7-9; vol. 4 (1974) 194-96. See also G. Jennison, *Animals for Show and Pleasure in Ancient Rome* (Manchester 1937) 21; W.C. McDermott, 'The Ape in Antiquity', *Johns Hopkins University Studies in Archaeology* 27 (1938) 77-78, 95-100; and the earlier discussion by Simon in his edition, vol. II, xx-xxii. McDermott does not mention the Rhesus monkey by name in connection with Galen. By contrast, Singer in his translation, above, note 6, 240, n. 22, holds that although Galen 'preferred the Barbary ape... it is probable that he relied chiefly on the Rhesus monkey.' Singer assembles some persuasive anatomical evidence in support of this claim, but no anatomical description of Galen's can be exclusively applied to *Macaca mulatta* (cf. Hill, 1966, 9-10). On the other hand, E. Savage-Smith, 'Galen's Account of the Cranial Nerves and the Autonomic Nervous System', Part 1, *CM* 6 (1971), 77-98, at 79, states that Galen 'did not use the Rhesus monkey but rather the then plentiful Barbary ape.' On the distribution and type of the *Macaca* species, see C. G. Hartman and W-L. Straus, *The Anatomy of the Rhesus Monkey* (London 1933).

16 Cf. Hill, *Primates*, above, note 15, 1974, 195.

17 Cf. ibid., 1970, 7-9.

18 For this last see *De usu part.* III 844 K.

19 Aristotle, *HA* 502a 16-b26. He also mentions – but without a complete description – the κυνοκέφαλος and κῆβος. Cf. Jennison, *Animals*, above, note 15, 20-21, 127-29; and Hill, *Primates*, above, note 15, (1966), 9, 213, and (1974) 194-96.

removed from the nature of man'.[20] According to Galen in Book XI of *Anatomical procedures*, this classification was known to the older anatomists.[21] The classes comprise (i) apes (a parody – μίμημα γελοῖον – of humans)[22] and ape-like animals; (ii) bears; (iii) pigs; (iv) saw-toothed animals; (v) horned two-hoofed ruminants;[23] (vi) hornless, smooth-hoofed animals.[24] Such a classificatory system clearly gave Galen enormous leeway, not only in what he could dissect but in enabling him to claim that the anatomical findings made from such animals could validly be applied to humans.[25] For the detailed dissections of the brain Galen made extensive use of the ox (*Bos taurus*), an ungulate, a member of his fifth class. One reason for this choice is given by Galen at the beginning of the discussion of the brain in the Greek part of Book IX, when he specifically mentions that the brains he is dissecting are ox brains, which, in the large cities at least, were everywhere on sale.[26] Availability of dissection material is, obviously, crucial for an anatomist.[27] Such procurement implies a steady and reliable source of availability: when Galen mentions that a dissector should be prepared to dissect other animals if there is a shortage of apes, for example, he may be acknowledging that members of this particular class were unavailable more often than not.[28] Apes, especially the Barbary ape, although known in the Roman Empire, may have been hard to come by.[29] Galen's own instructions (II 423 K.) to drown and not strangle the ape in order to avoid injury to the neck might also imply that the ape is a specimen that one does not readily encounter on a daily basis and thus all due care should be taken to see that no part of it is unnecessarily damaged.[30] In such a friable and quickly decomposing material as the brain, the ready procurement of dissection specimens is above all else crucial, and once more indicates why ox brains were Galen's preferred subject matter.

20 II 423 K. I. Garofalo suggests that the six-class classification, although alluded to earlier, was instituted by Galen, 'The six classes of animals in Galen', in J. A. López-Férez, ed., *Galeno. Obra, Pensamiento, e Influencia* (Madrid 1993), 73-87, at 86.

21 XI.2; 72.

22 *De usu part.* IV 126 K. The ape is γελοῖος with respect to the hand (*De usu part.* III 80 K.) and the muscles of the leg (III 264 K.).

23 II 430 K.

24 II 430-31 K.

25 Garofalo, Six classes, above, note 20, 85-86, adduces three reasons for Galen's use of such a variety of animals, citing availability, the relative differences in size, and the importance such a wide range of dissections had for the purpose of teleological argumentation. Teleology may well explain a significant part of Galen's motivation, but the ready availability of some animals and their size are arguably the crucial factors.

26 II 708 K.

27 J. M. C. Toynbee, *Animals in Roman Life and Art* (London 1973) 166, noted that goats were 'relatively cheap and easily obtainable sacrificial victims.' Cf. apes (55-60); cattle (148-62). That certain animals such as these were kept in sufficient numbers for sacrificial purposes made possible their ready procurement for other uses. See also O. Keller, *Die Antike Tierwelt*, 2 vols, (Leipzig 1909), I, 336-37; 402-03 and fig. 140.

28 II 227 K. Cf. Garofalo, Six classes, above, note 20, 85.

29 Pliny, *Hist. Nat.* VI 35,184, gives evidence for a *cynocephalus* and a *sphingion* form of ape, as well as a reference to varieties of ape in general (cf. VIII 80,215-16). But there is nothing here to indicate domesticated monkeys were bred, as Jennison claims, *Animals*, above, note 15, 128.

30 II 845 K. The neck dissection of an ape is described in *De nerv. diss.* II 845 K., *Admin. anat.* XI.1, and *De usu part.* III 390-91 K.

The other critical factor is that of size. Of all the animals Galen dissected on a regular basis, the brain of the ox was by far the largest. Of the apes used, their brains, on average, weighed half as much as that of the bovine.[31] The question of *comparative* size is a critical factor in Galen's approach to dissections in general: some points of anatomy *are* better seen in larger animals than in smaller ones.[32] Galen dissected many animals of varying size, ranging from elephants (albeit an incomplete dissection) to insects.[33] Conversely, useful generalisations on certain anatomical points drawn from larger animals could be employed to verify what cannot readily be observed in smaller, related ones.[34] The larger size of the ox brain relative to that of the other animals Galen regularly dissected, combined with its ease of availability, made it his uncontested choice.

How then, did the ape serve Galen's investigation of the brain? Galen did not ignore the brain and adnexae of the ape. The way he utilised this resource is instructional. Galen employed the brain and skull of the ape to demonstrate a number of quite discrete structures and thereby to impart a specific set of practical instructions. In other words, Galen's use of the ape brain is deliberately restrictive.[35] It is Galen's intention, in *Anatomical procedures*, Book IX, to delineate:

> ... the method of dissecting the parts of the brain while it remains in its place in the animal body. The dissection is best made in apes, and among the apes in such a one as has a face rounded to the greatest extent possible amongst apes. For the apes with rounded faces are most like human beings.[36]

31 The average weight of the brain of *Homo sapiens sapiens* is 1350g; that of the ox, 470.0 g. The following may be added: Lemur (*Lemur mongoz*), 20.0g; Spider monkey (*Ateles geoffroyi*), 61.5g; Gibbon (*Hylobates sp.*), 93.0g; Mandrill (*Mandrillus sphinx*), 178.0g; Chimpanzee (*Pan troglodytes*), 364.0g; Domestic hog (*Sus scrofa*), 98.0g. See S. Igarashi and T. Kamiya, *Atlas of the Vertebrate Brain* (Tokyo 1972). Although not all the species of primates Galen dissected are cited in this atlas, the lemur and mandrill have approximately similar brain weights to that of the Barbary ape. Only the brain of the young adult male of the Lowland Gorilla (*G. gorilla gorilla*), at 440.0g, is comparable to the ox brain. There is no evidence that this species of ape was ever brought to Rome (strictly, no primate is truly indigenous to the Mediterranean).

32 In studying the branch patterns of the abdominal aorta, Galen reflects that it is in *larger* animals that some parts of the body are better seen, *Admin. anat.* XIII. 8; 171. Cf. J. Scarborough, 'Galen's Dissection of the Elephant', *Koroth* 8 (1985) 123-34 (124).

33 Galen states that he dissected not just those animals belonging to the six classes, 'but also animals of the kind which crawl, those which move forwards by bringing the abdomen to their aid, water animals, and those which fly. And if I complete this work that I have started, as is my intention here, I want to dissect those animals also and to describe what there is to see in them.' *Admin. anat.* XI.12; 108. This project, which would have cemented Galen's reputation as the foremost comparative anatomist of Antiquity, was not fulfilled. Cf. Garofalo, Six classes, above, note 20, 85 n.68.

34 *Admin. anat.* XV.2; 227-28. Cf. F.J. Cole, *A History of Comparative Anatomy. From Aristotle to the Eighteenth Century* (London 1944) 47; Scarborough, The elephant, above, 32, 124 n. 6.

35 In Galen's use of the elephant, for example, it is not the *entire* animal Galen dissects; rather he examines its heart for two particular points of information: its similarity to the heart of other animals, and to solve the question of the presence or otherwise of the so-called heart bone, II 616-22 K. Cf. Scarborough, *The elephant*, above, note 32, 125.

36 IX.10; 10. This description fits πίθηκος, the Barbary ape, and σάτυρος (probably the Rhesus monkey). Cf. *De usu part.* III 844 K., where the ape which most resembles man is round-faced. Later in *Admin. anat.* (IX.11; 15) when Galen begins his account of brain vivisection, he advises against using an ape, in order to avoid the expression on its face during the procedure, recommending instead an animal such as a pig or goat, since they have the louder voices which the experiment demands. On the one hand, Galen could be hiding his anthropomorphic sensibilities in this demand for sustained vocalisation; on the other, he might also be reflecting the lack of regular availability of apes for vivisection.

An ape brain is thus to serve as a learning template for the student, enabling him to gain experience before embarking on the more detailed anatomical investigations of the (bovine) brain. This approach may also have the effect of partially lessening the tension between the use of the 'human-like' ape brain (the subject of superficial dissections only) and that of the ox (the focus of the more comprehensive anatomical research programme). But before proceeding in this manner with the brain of an ape, Galen, in *Anatomical procedures* Book IX, also enjoins the student to gain experience by performing dissections on the relevant areas of *other* animals:

> Should you have become practised previously in this mode of procedure on the carcase of a dead animal, then it will not prove difficult for you to carry it out well and correctly on a living animal, all around the greater part of the bones of the skull, without tearing away the dura mater along with the bone of the skull.[37]

The reference to using other animal specimens (here not named by Galen), underscores the anatomical value of the ape, and again implies that such specimens are not as readily available as one might suppose. That Galen stresses the need to find as close an approximation as possible to man is also obvious: Galen reminds his audience that the information obtained from such a source is directly applicable to that of man. This is crucial for understanding human osteology, where Galen shows some familiarity with the human skull. In *Anatomical procedures*, in discussing skull foramina, he remarked that:

> All these foramina you will see with your own eyes in a cadaver in which all that overlies the bones is decayed and the bones alone remain, in their connections with one another, without separating from each other. These can be seen in such human cadavers as you happen to look at ... and also in the bodies of apes when we have buried them for four months and more in earth that is not dry.[38]

The phrase, 'in such cadavers as you happen to look at', should not be taken as evidence that Galen had recourse to a human skull for daily study when composing his anatomical works (although it does not rule it out either). Nonetheless, an important part of Galen's own anatomical study in Alexandria involved the human skeleton, the only place where complete specimens were available for study.[39] The situation in Rome was different.[40] Of more immediate interest in the above citation is the mention of the ape skull. It is presented as an important source of osteological information and the careful preparation of an ape body by interment for four months underscores the value of such material.

37 IX.10; 13.

38 XIV.1; 182.

39 II 220 K. Cf. V. Nutton, 'Galen and Egypt', in Kollesch and Nickel, *Galen und das hellenistische Erbe*, above, note 13, 11-31 (15).

40 At *De comp. med. sec. gen.* XIII 604 K. Galen states that dissection was carried out on the bodies of the Germans in the Macromannic wars. Yet this was limited due, says Galen, to the physicians lack of anatomical knowledge. At II 385-386 K., Galen recommends taking advantage of the body of an exposed brigand. At II 386 K., Galen cites a 'frequent' practice of dissecting the bodies of exposed children, which had enabled anatomists to conclude that man had the same structure as an ape. This is cited in the third person plural, perhaps to draw attention away from Galen's own use of this material.

In *Anatomical procedures*, Galen proceeds with an exposition of the cranial bones in the ape, stressing their names, position, and sutures. This is done in order to stress their importance in the procedure of *trepanation* (ἀνάτρησις). Galen highlights this in the following way. After the entire skull of the ape has been exposed, the dissector is given meticulous instructions in locating and defining the sutures and their anatomical relationship with the underlying meninges (IX.10; 11). The relationship between the cranial bones and the outer meningeal layer (the *dura mater*) of the brain is linked by Galen to the arrangement of the cranial sutures. These serve as vital landmarks in trepanation in order to avoid contact with the dura. It was known by Galen's day that if the integrity of this membrane was compromised, the results were, more often than not, fatal. Trepanation was an established procedure for dealing with skull fracture, and of particular concern were the consequences of a *depressed fracture*, where there was danger of bone fragments pressing directly onto the outer meninx of the brain.[41] It is clear to Galen that such operations should be rehearsed many times until proficiency is gained. Moreover, to perform this procedure on a living animal enables the student to appreciate the importance of blood loss and to minimise trauma if possible. For Galen, the best way to gain such expertise is by utilising the skull bones and the dural relationships of an animal whose skull most resembles man. Thus he chooses 'apes with rounded faces'. They provide the best means for the most accurate application of a practical but potentially fatal technique. The skull and exposed brain of living apes therefore serve a different set of epistemological requirements compared to that of the bovine brain. They provide exact knowledge of vital anatomical landmarks for trepanation, and, secondly, they give the aspiring practitioner valuable preparation for later understanding the more detailed anatomy of the brain, since the key to such knowledge rests on experience, on learning the relationships between the skull, the membranous coverings of the brain and its underlying surface. The skull of the ape is for Galen a teaching tool, and one can well imagine that his students possessed a number of ape skulls.[42] Where there was neither a tradition of (nor, perhaps, the need for) anatomical illustration, the brain and especially the skull of an ape

41 The earliest Western use of trepanation for skull fractures is documented in the Hippocratic Corpus. In *Places in Man* trephining (using the more general word πρίειν) is recommended for what appears a depressed fracture, and the aim is to prevent the accumulation of purulent *fluid* within the wound (*Loc.*, 32.1; = E. Craik, *Hippocrates. Places in Man* (Oxford 1998) 70: 187-88). Three cases are described in *Epidemics V* (V 216, 227, 227-8 L). In the third case, the presence of exposed suture lines in the wound is an indication for prompt trepanation. A fourth case is given in *Epidemics VII* (V 404 L.). *On Wounds to the Head* cautions against too enthusiastic use of the trephine, even in cases of what are depressed and comminuted fractures, since there is a significant risk of damage to the dura with the use of the instrument (III 248-50; 258-60 L.). However, a fracture that is *not* comminuted is an indication for trephining (III 240-42 L.). Galen limits the use of trephining to the relief of pressure and its consequences; that is, to trauma involving the skull, and in such cases as the draining of phlegmatous lesions on the head. See J. Rocca, 'Galen on the uses of trepanation,' in *Trepanation – History, discovery, theory*, ed. R. Arnott et al. (Lisse 2002), 253-71.

42 Properly dried and prepared ape skull bones are also indispensable as aids in studying the nerve and vascular foramina, and the sutures. If this preparation is less than perfect, then, as Galen notes, errors of interpretation may arise: *Admin. anat.* XIV.1; 182-83. See also Garofalo, *Procedimenti,* above, note 7, III, 1038-39. On the use of an ape skull in tracing the intracranial course of the optic nerve, see *Admin. anat.* X.1; 28. Cf. X.6; 54 (the mandible); XIV.1; 182 (cranial nerve foramina and preparation of an ape skeleton); XIV.4; 196-197 (cranial nerve foramina and the importance of animal skulls in general as learning guides). In discussing the types of animals to be used in dissection of the orbit and its contents, Galen reiterates the importance of the ape skull as a learning template. *Admin. anat.* X.1; 29.

served both documentary and pedagogic purposes.[43] They comprise a set of source materials to which the student may repeatedly refer during the more detailed investigation of the brain and its nerves. Furthermore, such practical knowledge of the skull and its immediate relationships is crucial if experiments on the brain of a vivisected animal are to be performed.

III

Galen's experiments on the ventricles reflect the methodology that characterises his entire approach to the physiology of the brain and the nerves.[44] Furley and Wilkie conclude that Galen's '... experiments on the nervous system, in particular, carry total conviction as accounts of what he had himself actually done and seen.'[45] The experiments Galen performs on the ventricles of the brain are crucial for his conceptualisation of the brain as the organ of the rational soul. For him, the brain is the *hegemonikon* for voluntary motion and sense perception.[46] The ventricular system is the locus of this controlling centre.[47] *Anatomical procedures* Book IX tells how experiments may be performed on these cavities in order to establish their physiological status.[48] Above all, Galen's anatomical works are interactive manuals in which he maintains and demands an impressively high standard. The descriptions of his vivisectional experiments are designed to reflect and reinforce those criteria. For Galen, dissection enables an anatomist to move beyond mere *beliefs* about the structure and function of the living body. This forms the basis of Galen's attacks on certain of his peers, who, whilst being referred to as *anatomists*, nevertheless perform no careful or precise dissections. Indeed, if they regard dissection as unimportant, then it follows they will have no inkling of the possibilities of determining function in the living animal by vivisection. Why then should they bother, concludes Galen in a telling phrase, 'to cut or ligate' the living

43 There is no evidence that Galen resorted to pictorial representations of the brain in his demonstrations to students or colleagues. He does, it is true, make use of a diagram in his description of the insertion and origin of the deltoid muscle (II 273-74 K.) and the cervical part of the trapezius muscle (II 445-46 K.). These are geometric idealisations which are meant to aid, not substitute for dissection. Cf. R. Herrlinger, *History of Medical Illustrations from Antiquity to 1600* (London 1970) 9-24.

44 For an overview see J. Rocca, 'Galen and Greek Neuroscience (notes towards a preliminary survey)', *Early Science and Medicine* 3 (1998) 216-40. I provide a more thorough study in my *Galen on the Brain: anatomical knowledge and physiological speculation in the second century AD* (Leiden forthcoming). Cf. also T. Manzoni, *Il cervello secondo Galeno* (Ancona 2001) 9-38.

45 D. J. Furley and J. S. Wilkie, *Galen On Respiration and the Arteries* (Princeton 1984) 48: 'If we deny this we have to assume a degree of duplicity on the part of Galen that seems totally incredible.' Cf. J. Prendergast, 'The Background of Galen's Life and Activities, and its Influence on this Achievements', *Proc. Roy. Soc. Med.* 23 (1930) 1131-48. For important background studies on Galen as an experimenter see A. Debru, 'L'expérimentation chez Galien', *ANRW* II. 37.2, 1994, 1718-56; M. Grmek, *Il calderone di Medea: La sperimentazione sul vivente nell'Antichità* (Rome 1996) 101-22.

46 Galen never addresses the broader question of in what way dissection of irrational animals can demonstrate that the human brain is the *hegemonikon* of the rational soul.

47 J. Rocca, 'Galen and the Ventricular System', *J. Hist. Neurosciences* 6 (1997) 227-39.

48 A more extensive discussion is found in *De plac. Hipp. et Plat.* VII: *CMG* V.4.1.2, 442-44. There are also important references to pathological effects on the ventricular system in *De usu part.* (III 663-65 K.) and *De locis affectis*, VIII 128, 230-31, 232-3, 270 K.

animal?[49] The Galen encountered setting down his observations is someone exceptionally well-trained in a set of techniques which makes vivisection not only possible but consistently repeatable in his hands.[50]

Nowhere is Galen's technical proficiency better seen than in his ventricular experiments. In order both to expose the ventricles and to keep the subject alive, Galen must perform delicate and sophisticated surgery. Such procedures represent major operations on the brain.[51] Galen recommends that 'for this purpose you must procure either a pig or a goat, in order to combine two requirements. In the first place, you avoid seeing the unpleasing expression of the ape when it is being vivisected. The other reason is that the animal on which the dissection takes place should cry out with a really loud voice, a thing one does not find with apes'.[52] Exposing the brain of a living animal clearly requires, at the very least, the same techniques refined upon the dead. The crucial difference between dissection and vivisection is that, in the latter, the skilled operator must also be prepared to deal with blood loss.[53] That the animals survived the post-operative period for sufficient time for Galen to record any meaningful results is tribute to his operative skill and to his development of a set of techniques that minimised trauma and blood loss. With the top of the skull removed, haemostasis achieved in the living animal, and the dura mater opened (IX.12; 17-18), each step again emphasising Galen's technical proficiency, the brain is now exposed, and the animal examined for any adverse effects from the procedures so far executed:

> For the meninx as I described it is in all parts separable from it, as I have shown, except at the sides of the sutural lines which I indicated, where the meninx becomes folded into two layers as described, and sinks downwards penetrating a considerable distance further into the parts of the brain lying beneath it [*falx cerebri* and *tentorium cerebelli*] ... Now you can cut away these three parts of the meninx [*dura mater*] and thus expose the underlying portions of the brain. Two of these three parts lie in the region beneath the parietal bone, the third is the portion overlying the hind brain (*cerebellum*). When you have done that, then make an inspection and ascertain for yourself whether the animal is being

49 II 232 K. But Galen's scornful language does not show that they avoided all dissection, or even vivisection, but merely that they did not carry it out as he would have done. For one of his anatomical opponents, see above, p. 14.

50 Cf. N. Mani, 'Die wissenschaftlichen Grundlagen der Chirurgie bei Galen (mit besonderer Berücksichtigung der *MM*)', in *Galen's Method of Healing*, ed. F. Kudlien and R. J. Durling (Leiden 1991) 26-49. Galen's experience in surgery as physician to the gladiators at Pergamum (157-161 AD) provided useful experience in dealing with wounds in living subjects, as well as valuable observations of patients' behaviour in wounds to the head. His anatomical skills, claimed Galen, won him this position at Pergamum. Cf. *De optimo medico cognoscendo*, *CMG* Suppl. Or. IV, 103.10-105.19. L. H. Toledo-Pereyra, 'Galen's Contribution to Surgery', *JHM* 28 (1973) 357-75, claims to offer a summary of Galen's surgical skills, but draws many of his examples from the pseudo-Galenic *Introductio sive medicus*.

51 There is no evidence that Galen undertook his later experiments on the ventricular system, 'to verify the conclusions which he had already drawn from surgical experience', as R. E. Siegel thought, *Galen on Psychology, Psychopathology, and Function and Diseases of the Nervous System. An analysis of his doctrines, observations and experiments* (Basel 1973) 43.

52 IX.11; 15.

53 As Galen puts it, 'nothing so disturbs any surgical procedure as haemorrhage', II 628 K. See also *Admin. anat.* IX.11; 15-16, where Galen describes the minimisation of blood loss when the pericranium is incised and retracted. On Galen's haemostatic methods, see G. Majno, *The Healing Hand. Man and Wound in the Ancient World* (Cambridge, Mass. 1975) 403-04; C. F. Salazar, *The Treatment of War Wounds in Graeco-Roman Antiquity* (Leiden 2000) 43-44.

deprived of respiration, voice, movement or sensation, or whether none of these defects is showing itself in it, either at the time when the incision was made upon it or else soon afterwards. The latter may quite well be the case, when it happens that the air is warm.[54]

Galen is seeking to establish that, with the brain substance untouched, the animal's behaviour is unchanged. In the last sentence, Galen also implies that deleterious effects will not necessarily occur immediately, but 'soon afterwards' if the air is warm. The corollary is that cold air either increases the severity of distress shown by the animal, or else retards its recovery.[55] Galen clarifies this by noting his observations on the effects of external temperature upon the exposed living brain; again, Galen attempts to minimise the number of outside influences on the brain before ventricular experimentation begins:

But if the air is cold, then in a degree corresponding to the amount of the cold air streaming in upon the brain, each single one of these functions of the brain that we have mentioned weakens; the animal remains for a certain length of time unconscious, and then expires. Therefore it is best that you should take in hand the detachment of the dura mater from the skull in the summer time, or, if you perform it at another season, no matter which that season may be, you should heat the room in which you intend to dissect the animal, and warm the air.[56]

Once the brain of a living animal has been exposed, Galen documents the nature of the experiments he performs on its ventricles. It is worth quoting this passage in its entirety:

Should the dissection be thus performed, then after you have laid open the brain, and divested it of the dura mater, you can first of all press down upon the brain on each of its four ventricles, and observe what derangements have afflicted the animal. I will describe to you what is always to be seen when you make this dissection, and also before it, where the skull has been perforated, as soon as one presses upon the brain with the instrument which the ancients call the 'protector of the dura mater'.[57] Should the brain be compressed on both the two anterior ventricles, then the degree of stupor which overcomes the animal is slight. Should it be compressed on the middle ventricle, then the stupor of the animal is heavier. And when one presses down upon that ventricle which is found in the part of the brain lying at the nape of the neck (the *fourth* or *posterior ventricle*), then the animal falls into a very heavy and pronounced stupor. This is what happens also when you cut into the cerebral ventricles, except that if you cut into these ventricles, the animal does not revert to its natural condition as it does when you press upon them. Nevertheless it does sometimes do this if the incision should become united.[58] This return to the normal condition follows more easily and more quickly, should the

54 IX.12; 18. I have added the final gloss [Duckworth].

55 Cf. *De loc. aff.* VIII 161-62 K., where cold affecting the brain produces torpor and somnolence.

56 IX.12; 18.

57 This is also known as the 'protector of the meninx', the *meningophylax* or *membranae custos* (Celsus, *Med.* VIII.3). Galen here takes the opportunity to note similar effects from another procedure, trepanation. Galen refers to this instrument in a way that strongly indicates that it was something with which he was thoroughly familiar (cf. *In Hipp. Epid. comm.* III.1: *CMG* V.10.2.1, 25).

58 Galen is deducing wound closure due to tamponade, where the brain substance overlying the ventricles seals the wound.

> incision be made upon the two anterior ventricles. But if the incision encounters the middle ventricle, then the return to the normal comes to pass less easily and speedily.[59] And if the incision should have been imposed upon the fourth, that is, the posterior ventricle, then the animal seldom returns to its natural condition; although nevertheless if the incision should be made into this fourth ventricle, provided that you do not make the cut very extensive, that you proceed quickly, and that in the compression of the wound in some way or other you employ a certain amount of haste, the animal will revert to its normal state, since the pressure upon the wound is then temporary only – and indeed especially in those regions where no portion of the brain overlies this ventricle, but where the meninx only is found. You then see how the animal blinks with its eyes, especially when you bring some object near to the eyes, even when you have exposed to view the posterior ventricle. Should you go towards the animal while it is in this condition, and should you press upon some one part of the two anterior ventricles, no matter which part it may be, in the place where as I stated the root of the two optic nerves lies, thereupon the animal ceases to blink with its two eyes, even when you bring some object near to the pupils, and the whole appearance of the eye on the side on which lies the ventricle of the brain upon which you are pressing becomes like the eyes of blind men.[60]

The above passage is without doubt one of the most impressive accounts of physiological experimentation extant in Western Antiquity. The claims made are far reaching and cannot be fully entered into here. But whether Galen deals with the effects of pressure or incision, the results are remarkable. His techniques are able to account for the function of each ventricle. This account should be read as a composite of Galen's brain investigations, since clearly they cannot have been the results of any single experiment. He presents his readers and students with a formidable – even overwhelming – combination of factors that few could manipulate successfully, let alone concurrently. Underlying this experimental *tour de force* is the message that failure to observe what Galen has expounded means only that the procedure has been improperly carried out; not that the methodology or the results can be called into question. These experimental and polemic techniques form the mainstay of Galen's experimental methodology of the brain.[61] The rest of the brain is deliberately excluded from having a bearing on these experiments:

> ... the animal, when one pierces or incises the thin meninx, sustains no derangement as a result, just as none such befalls it if the brain should be incised without the incision reaching as far as to one of its ventricles.[62]

59 Galen does not formally quantify his observations, beyond noting the presence or absence of his key parameters of motion and sensation. An ambiguous estimation of the time taken to return to a normal state is as far as Galen goes in terms of attempting to quantify his results. On quantification in Antiquity see J. J. Bylebyl, 'Nutrition, Quantification and Circulation', *BHM* 51 (1977) 369-85; Grmek, *Il calderone di Medea*, above, note 45; O. Temkin, 'A Galenic model for Quantitative Physiological Reasoning?', *BHM* 35 (1961) 470-75.

60 IX.12; 18-19. My gloss in parentheses.

61 Apart from incision and pressure, the other experimental technique Galen employed on the brain was that of *ligation* used principally to determine nerve function.

62 *Admin. Anat.* IX.12; 20. Cf. *De plac. Hipp. et Plat. CMG* V 4,2, 446,27-29.

In *Anatomical Procedures*, Galen also serves notice that his ventricular experiments may further be manipulated to produce a satisfactory outcome – a small incision, performed quickly and with prompt compression applied to the wound may result in the animal returning to its normal state.[63] To stress that it is the ventricle, not the overlying brain substance, that determines function, Galen, later in Book IX, reiterates the consequences of incising the region of the posterior ventricle:

> ... that should the spinal marrow that lies between the skull and the first vertebra be severed, or the meninx which protects the end of the posterior ventricle of the brain be cut through, then at once the whole body of the animal becomes deprived of movement. It is just here that you will see, in the temples of the gods, the oxen receive the stab when the so-called sacrificers of oxen cut into them.[64]

Apart from experimental evidence, animal sacrifices yield for Galen important clues. The instantaneous death of oxen that occurs when the sacrificial cut is made at the level of the first cervical vertebra, is interpreted by Galen in terms of the topography of the posterior ventricle and its pneumatic contents, as the region incised is precisely where he maintains the posterior ventricle ends and the spinal cord begins.[65] Erasistratus observed, as did Galen, that damage to the brain in the area of the first cervical vertebra, usually resulted in death. As that part so incised is indeed covered by dura mater, Erasistratus reasoned that death was due to damage to the dural meninx and made the valid empirical deduction that the dura was thereby responsible for nervous transmission in some way.[66] To Galen, Erasistratus is almost correct; but the importance of damage to the meninx for Galen is that it exposes the posterior ventricle and its contents, as the posterior ventricle (according to him) empties via a single channel into the beginning of the spinal medulla.[67] Galen validates his interpretation of these observations, and dismisses the Erasistratean thesis, by pointing out that damage to the meninx anywhere else does not result in the animal becoming motionless.[68]

If the effects of incising the ventricles are broadly similar to those elicited by pressure, it may be asked why Galen performs them. The answer lies in the nature of the contents of the ventricles. Ventricular incision is the first of a two-stage experimental process, the second of which creates the conditions for resealing the ventricle, allowing the animal to recuperate, and then observing the results in each case. These results are interpreted by Galen in *pneumatic* terms. That the animal recovers is ascribed to the replenishment of *psychic pneuma* following closure of the incised ventricle. *Pneuma*, however, is not mentioned in *Anatomical Procedures*, although, like Banquo's ghost, its presence may be felt. For

63 As expressed in *De plac. Hipp. et Plat. CMG* V 4,2, 442.30, an incision to the posterior (fourth) ventricle causes the most harm, with damage to the middle ventricle being the next most severe.

64 *Admin. anat.* IX.14; 25.

65 Cf. *Admin. anat.* IX.14; 25. It is well known to butchers that, 'the medulla may be cut by passing a knife between the skull and the atlas, less easily, between the atlas and axis.' R. Macgregor, *The Structure of the Meat Animals. A guide to their anatomy and physiology*, 3rd edn. Revised by F. Gerrard (Oxford 1980) 99. That is, between the skull and the first cervical vertebra (in the first instance) and between the first and second cervical vertebra (in the second).

66 I. Garofalo, *Erasistrati Fragmenta* (Pisa 1988) Fr. 42A. Cf. E. D. Phillips, *Greek Medicine* (London 1973) 148.

67 *De plac. Hipp. et Plat, CMG* V 4,2, 446,20-22.

68 An Erasistratean might well object that this proves nothing; the meninx may well be at its most vulnerable in this area, and damage to it will have an inevitably fatal outcome.

pneuma's role in these events, one must look elsewhere. As we have seen, above, Galen had noted that an incision about the posterior ventricle renders the animal instantly motionless. It is the escape of pneuma that is responsible for this state. As he put it in *On tremor*, no other substance is as capable of *emptying* or *collecting* again so easily. Pneuma is also capable of moving into the body *instantaneously*.[69] This can be used to explain why the animal recovers when the incision has been sealed. For Galen, only pneuma accounts for the results of ventricular incision.[70] His experiments indicate that damage to the brain – interpreted in terms of impairment affecting the ventricular system – produces a range of *affective states* (παθήματα) from reversible stupor to death. Galen creates a scale of function based on relative and absolute incapacity as recorded by experiment. It is the integrity of the ventricular system that is of first importance in maintaining the brain's role as *hegemonikon* of the rational soul. Contained within the ventricles, the position of psychic pneuma, however, is more problematic. Galen has several different theories for pneumatic action, but its precise manner of operation is something Galen never entirely resolves.[71]

IV

This discussion has shown something of the comprehensive nature of Galen's edifice of the anatomy of the brain and the physiological argumentation he employs to establish its hegemonic status. Galen makes effective use of his anatomical materials. The head of the ape that is *most like man* is unrivalled as a learning template. The internal anatomy of the ape brain, is, however, another matter.[72] In any case, Galen does not need to rely on the brain of an ape for detailed dissections of the structure and substance of the brain. He has a larger, more readily available subject for his studies in the ungulate, and it is this class of animal which provides the model for his construction of the anatomy of the human brain. Within that template, Galen provides a meticulously detailed account of ventricular anatomy and function. Here, of course, one is dependent both on the authority of Galen's descriptions and his experimental accounts. To be sure, only a skilled vivisector may with any confidence manipulate the subject material in such a way in order to exclude the meninx or the brain substance from having any bearing on the experiments. Such, at any rate, is Galen's intention and whilst his experimental methodology is internally consistent, the physiology behind it is necessarily speculative. However, since Galen places *psychic pneuma*, his chosen effector agent of the soul, within the ventricular cavities, he needs to show that the brain is the hegemonic organ of sensation and voluntary motion by experimenting on those parts of it

69 *De tremore, palp, convuls. et rig.* VII 596-97 K. The context is the role of pneuma as the cause of palpitation. At *De plac. Hipp. et Plat. CMG* V 4,2, 416.5-7, the heart is said to be the origin of the 'pneuma-like and boiling blood'.

70 Galen spells this out this relationship between the ventricle and its contents in *PHP*: *CMG* V 4,2, 446,11-17. Although this is a weak claim for pneuma *per se*, Galen ensures by this discussion that the argument for hegemonic ventricular function returns to its anatomical base.

71 See J. Rocca, 'Galen and Greek Pneuma Theory. The Limitations of Physiological Explanation', in *Philosophy and Medicine. II. Studies in Greek Philosophy 29*, ed. K. J. Boudouris (Athens 1998) 171-97.

72 This lack of attention to the base of the skull may further be inferred because Galen never provides an explicit citing of the lack of the *retiform plexus* in any dissection involving the brain of an ape. None of the anatomical descriptions Galen gives of the retiform plexus can be found to match conclusively any arterial structure of the base of the brain of an ape, whereas they point unequivocally to the ungulate.

amenable to his particular skills. It is therefore essential that the results from such experiments are comprehensible to Galen and to his audience: in other words, that such observations may be directly related to function. To this end, Galen devotes considerable effort in seeking to establish that the ventricles of the brain possess a set of distinct characteristics which eminently fit that organ for hegemonic status, and no other.

To quote once more from *Anatomical procedures* (IX.7; 4), Galen states that it is:

> ... not here my purpose to derive the knowledge of the nature of the things which I wish to understand from analogy; for this is not the aim of anatomy. Rather I am simply trying to give an account of those things which manifest themselves to the eyesight.

This is a typically Galenic statement. His anatomical epistemology and the strategies which underpin it, are, of course, far more complex than what he ostensibly maintains. Book IX of Hunayn's text at least allows us to see some of the ways in which he applies that knowledge. Refuting some and dazzling others, brain anatomy in Galen's hands is a powerful if selective tool.

University of Cambridge and the Karolinska Institute, Stockholm.

THE PATHOLOGY OF PREGNANCY IN GALEN'S COMMENTARIES ON THE *EPIDEMICS*

REBECCA FLEMMING

The relationship between Galen and Hippocrates is a crucial one for both parties. For Galen the 'divine Hippocrates' was a key figure in the construction of his own medical authority, as he laid claim to be the true heir of the founding father of Greek medicine. And, whatever the justice of this assertion, the Pergamene certainly proved to be the most historically successful claimant of the title: it is his version of the Hippocratic tradition, his interpretation of Hippocratic thinking, which has transmitted itself most effectively down the ages.[1]

It is the most concentrated form of this Galenic rendition of Hippocratic ideals and doctrines which is of concern here, that is his actual commentaries on Hippocratic texts. Wesley Smith dates the first of these – his commentaries on the main surgical works ascribed to Hippocrates – to about AD 175, that is to Galen's 'full maturity'; and sees it as a move to justify his repeated, but up till then little substantiated, assertions of Hippocratic filiation.[2] Galen himself states that his exegetical endeavours were all undertaken at the urging of friends – though this is a commonplace in his oeuvre – and reveals, rather more implicitly, that tradition too urged him in this direction: elucidating Hippocratic texts had become a standard part of the literary repertoire of Greek medicine from the Hellenistic period onwards.[3] It is clear that the Hippocratic Corpus was a key resource for all those situated in the more intellectually ambitious sectors of classical medicine; a complex legitimating charter which both served to underwrite the medical enterprise in general, and could be mobilised to support more specific projects. Not all cast themselves as Hippocratics – Asclepiades of Bithynia and many Methodists, for example, seem to have ploughed a more explicitly innovative furrow – and not all attempted to shape a Hippocrates in their own image; but this was an acknowledged avenue to medical authority, and even those who disdained it did not ignore the Hippocratic writings. Asclepiades, for instance, wrote at least two Hippocratic commentaries himself.[4]

1 The most extensive and definitive discussion of relations between Galen and Hippocrates is in Wesley D. Smith, *The Hippocratic Tradition* (Ithaca, N.Y. 1979).

2 Smith, *Hippocratic Tradition*, esp. 122-25.

3 For his most extensive appeal to these friendly exhortations see the introduction to the second part of Galen's commentary on *Epidemics 3* (*CMG* V 5.10.2.1, 60.4-63.9). For an outline of Hippocratic scholarship before Galen see Smith, *Hippocratic Tradition,* 177-246.

4 Caelius Aurelianus refers to his commentary on *Aphorisms* (*CP* 3.5), and Galen engages with his commentary on *In the Surgery* (18B 666, 715, 805, 810 K). Smith discusses Asclepiades' attitude to Hippocrates at *The Hippocratic Tradition*, 222-26.

Galen certainly pursued the Hippocratic path to success. So there was a lot at stake in his commentaries, especially since they were undertaken *after* he had outlined a system with which his Hippocrates had to conform; a system, indeed, that was in large part secured by that claim to conformity. What this paper seeks to do, therefore, is explore the tensions inherent in Galen's exegetical project through a close examination of certain sections of this project. I have selected for this scrutiny, Galen's treatment of a particular set of female case histories in the *Epidemics*, that is those in which disease, death, and cure are most closely connected with the reproductive process, especially pregnancy and birth; topics about which Galen is otherwise somewhat reticent. I hope, therefore, both to illuminate Galen's activity as a Hippocratic commentator and to investigate specific aspects of his approach to woman as an object of medical knowledge.

About a third of the patients who appear in the seven books of Hippocratic *Epidemics* are women, and many of them have fallen ill after childbirth or miscarriage.[5] This, in itself, raises issues about the doctrinal relationship between the various books of *Epidemics* and other works in the Hippocratic Corpus, particularly those which peddle their pro-natalist line most strongly. For, as scholars have noted, Hippocratic gynaecology, as most programmatically represented by treatises such as *Diseases of Women*, assumes and asserts the dependence of female health on reproduction function.[6] The female body was loose and spongy in texture, prone to the accumulation of surfeit liquids, which required evacuation – through menstruation, or ideally childbearing (the surfeit then going to nourish the foetus) – in order to rectify its imbalances. Other benefits accrued from regular intercourse, pregnancy and birth, as this kept the womb moist and stable; but the bottom line for health was balance, which the well-purged, reproductive woman maintained, and the non-menstruating, non-procreative woman dangerously lacked. Hence the therapeutic mantra that, 'if she becomes pregnant she will be healthy', so often repeated in *Diseases of Women*, sometimes preceded by the more practical injunction, 'have her go to her husband.'[7]

However, as Ann Ellis Hanson has shown, there is in fact a substantial overlap of basic ideas and understandings between the *Epidemics* and Hippocratic gynaecology.[8] Moreover, though there is a fairly high attrition rate following birth (and miscarriage) in several books of the *Epidemics* (there is a high attrition rate in general in these works), that is only to be expected in a set of texts which disdain the normal and healthy in favour of documenting

5 Nancy Demand, *Birth, Death, and Motherhood in Classical Greece* (Baltimore and London 1994), 167 gives the percentage of female cases in each book as: 29% in Book 1, 55% in Book 2, 43% in Book 3, 36% in Book 4, 26% in Book 5, 32% in Book 6, and 25% in Book 7. She also gives a further breakdown of the percentage of cases which are pregnancy related. The second appendix (168-183) contains translations of all the cases involving pregnancy.

6 See e.g. Ann Ellis Hanson, 'The medical writers' woman', in D. M. Halperin, J. J. Winkler, and F. I. Zeitlin (eds.), *Before Sexuality* (Princeton 1990), 309-38. On *Diseases of Women* itself see also Hermann Grensemann, *Hippokratische Gynäkologie: Die gynäkologischen Texte des Autors C nach den pseudoHippokratischen Schriften de Mulieribus I, II und de Sterilibus* (Wiesbaden 1982).

7 See e.g. Hp., *Mul.* 1.37, 2.131 and 2.135 (8.92.7-8; 280.2-3 and 308.2-3 L) for the combination.

8 Ann Ellis Hanson, 'Diseases of women in the Epidemics', in G. Baader and R. Winau (eds.), *Die Hippokratischen Epidemien: Theorie-Praxis-Tradition* (Stuttgart 1989), 38-51; and for a wider discussion of relations between *Epidemics* and other texts in the Corpus see e.g. Jacques Jouanna, 'Place des Epidémies dans la Collection hippocratique: le critère de la terminologie', in the same volume, 60-87.

what can, and has, gone wrong with various human bodies.[9] It is worth noting that the *Epidemics* contain only one case of death occurring during pregnancy itself (without miscarriage) – that is the case of the wife of Antimachus – all the others who fall sick during pregnancy recover.[10] Furthermore, the notion of therapeutic procreation still persists. The wife of Gorgias has long-standing suppression of the menses and uterine pains which are cured after a rather complicated and difficult double pregnancy followed by one birth and a miscarriage.[11] Two unnamed women in Book Two are cured, at least temporarily, of strangury and hip pain respectively by pregnancy or birth.[12] It also has to be said that reproductive dangers clearly co-exist with strong encouragements to maternity even in *Diseases of Women.* Before reaching the more concertedly therapeutic sections in which the equivalence of pregnancy and health begins to be repeatedly stressed, there is a comprehensive catalogue of what can go wrong with each phase of the female procreative (and regulative) pattern, from menstruation to birth, and its aftermath. The interdependence of female well-being and reproduction (which encompasses menstruation) is thus illustrated from both sides; for it is precisely in all these processes that the difference between well-being and illness primarily resides, and they can go either way. There is some reassurance even here, however, as the list of problems both opens and closes with the assertion that it is women who have never given birth who are most susceptible to these misfortunes.

Galen comments on only four books of the *Epidemics* – Books One to Three, and Book Six. Moreover, these commentaries were produced over about a decade, and are quite divergent in style and approach. As Smith describes them, an alternating pattern may be discerned.[13] The commentaries on Books One and Three are relatively brief, consisting basically of paraphrase and explanation in Galen's own terms, and revealing little of the long exegetical tradition which he follows. Galen's treatment of Books Two and Six, on the other hand, is much fuller: more expansive in its discussion of the lemmata, and more extensive in its engagement with previous commentators. This contrast can, in part, be related to the divergent nature of the texts themselves: the polished literary qualities of *Epidemics* 1 and 3 are obvious, while 2 and 6 are much less finished products – heterogeneous and cryptic.[14] Commentary, therefore, collected more thickly around these more problematic texts: attracted by disputes about authorship, and the sheer difficulty (but also interest) of their formulations. Galen was no exception, and while holding that *Epidemics* 2 and 6 are posthumous productions, fashioned out of Hippocrates' notes by a literary executor such as his son Thessalus, he clearly found them intellectually fascinating. Galen's own contributions to this tradition of explication have also had mixed fortunes in terms of their transmission – the Greek text of his commentary on *Epidemics* 2 published in Karl Gottlob Kühn's monumental

9 Demand notes (*Birth*, 44) that women with diseases relating to pregnancy in *Epidemics 1* and *3* suffer a staggering 80% mortality rate, as against a death rate of 54% for the rest. However, in Books 2, 4 and 6, no pregnancy related fatalities are recorded.

10 Hp. *Epid.* 5.18 (5.216.20-218.13 L).

11 Hp. *Epid.* 5.11 (5.210.12-212.4 L).

12 Hp. *Epid.* 2.2.17 and 18 (5.90.7-92.2 L).

13 Smith, *Hippocratic Tradition*, 123-67.

14 For general discussion of the differences between the books of *Epidemics* see Wesley D. Smith, 'Generic form in Epidemics I to VIII', in Baader and Winau (eds), *Die Hippokratische Epidemien* (1989), 144-58.

edition of Galen's work (1821-3) is a Renaissance forgery, and it really only survives in an Arabic translation, as do various other sections of the works, such as the final parts of the commentary on *Epidemics 6*.[15] Improved versions of the Greek, together with Franz Pfaff's German translations of the surviving Arabic portions (though unfortunately not the Arabic texts themselves) are now to be found in the *Corpus Medicorum Graecorum* series.[16]

Despite his selectivity, Galen must still confront a plethora of female case histories in his commentaries, many more than appear in his own, original, writings, and with a different, more procreative, emphasis. Galen's extensive discussions of reproduction are concentrated in his anatomical and physiological works, where concrete cases are rare.[17] Nor does he work with any distinctly gynaecological or obstetrical categories of disease, which would need to be exemplified in his pathological and therapeutic treatises. Miscarriage, and other reproductive difficulties, certainly appear on various lists of the possible causes of disease (and they may also be caused by other ailments), but they cause (and are caused by) generic illnesses, like fever, not anything requiring more particular treatment. The generality of the issues involved is perhaps best illustrated by the fact that Galen's only substantial case history dealing with miscarriage comes in his work *On Examinations of the Best Physicians*, and is there to make a point about the crass stupidity of the woman's husband and his signal inability to recognise the best physician – that is, of course, Galen – even when he is displaying his diagnostic brilliance right in front of his eyes.[18] The complicated miscarriage is, except in its very complications, which provide Galen with the opportunity to so outshine his rivals, entirely incidental to the story. Hence the second angle of interest in Galen's discussion of these female case-histories in the *Epidemics*: how does he deal with these cases of whole, actually procreative, women for whom something has gone wrong? How does he grapple with a category he found uninteresting, except as placed before him by the divine Hippocrates?

The first, and most important, move Galen makes in relation to these cases is to explain them. Explication and elaboration of the bald (and often enigmatic) descriptions and assertions that make up many Hippocratic treatises is what Galen's commentaries are most consistently about; explication within the terms of his own medical system, inspired but not constrained by Hippocratic notions as it is. His treatment of the women in the *Epidemics* is no exception, and it is with reference to the Thasian woman by the cold water from *Epidemics 3* that Galen gives the fullest explanation for why so many of these women fall ill (and often die) after birth. The same understanding is implied, or offered more partially, in all the other cases, and is sometimes even applied more widely. So it is worth quoting the relevant passages in full.

The Galenic lemma provides a somewhat abbreviated version of the original:

> Δεύτερος ἄρρωστος. ἐν Θάσῳ τὴν κατακειμένην παρὰ τὸ ψυχρὸν ὕδωρ ἐκ τόκου θυγατέρα τεκοῦσαν καθάρσιος οὐ γενομένης πυρετὸς ὀξύς, φρικώδης τριταίην ἔλαβεν, καὶ τἄλλ<α τὰ> ἐφεξῆς ἄχρι τῆς ἀρχῆς τοῦ τρίτου ἀρρώστου.

15 On Kühn and his edition, see Vivian Nutton's essay in this volume, 1-7.

16 *CMG* V 10.1-v.10.2.4 (1934-1960).

17 For an extensive discussion of Galen's treatment of women in his works, covering all the points made here, see Rebecca Flemming, *Medicine and the Making of Roman Women* (Oxford 2000), 288-358.

18 Gal. *Opt. Med. Cogn.* 13.6-8 (*CMG Supp. Or.* iv 130.13-132.11).

> [Second illness] In Thasos, the woman who lay sick by the cold water, on the third day after giving birth to a daughter, not having purged, was seized with an acute fever accompanied by shivering; and the rest, in order, until the beginning of the third illness.[19]

The commentary begins with a gratuitous dig at Sabinus, one of Galen's chief rivals as Hippocratic interpreter, and then moves onto the explanatory core of his exegesis:

> φαίνεται δ'οὖν ἐπισχεθείσης τῆς μετὰ τὸν τόκον καθάρσεως ἡ νόσος γενέσθαι τῇ γυναικί. νοσώδης μὲν γὰρ καὶ ἡ τῶν καταμηνίων ἐπίσχεσις, ἀλλ' οὐχ ὁμοίως βλαβερὰ τῇ μετὰ τὸν τόκον, ὅτι μὴ μόνον αὕτη πλῆθος ἀλλὰ καὶ κακοχυμίαν ἱκανὴν ἐργάζεται. τὸ μὲν γὰρ χρηστότερον αἷμα τὸ ἔμβρυον ἕλκον ἑαυτὸ τροφῆς ἕνεκα, κατάλοιπον δὲ τὸ φαυλότερον αἴτιον γίγνεται τῆς κακοχυμίας ταῖς κυούσαις, ἣν μετὰ τόκον ἡ φύσις ἐκκενοῖ.

> For it seems that the disease was engendered in the woman by the retention of the post-partum purge. For the retention of the menses tends to produce disease, but is not as damaging to the woman as [retention] after birth, since not only is [this retention] itself a excess, but it also produces considerable cacochymy (i.e. evil humours). For the embryo attracts the most useful blood to itself, as nourishment, and the poorer remainder becomes the cause of cacochymy in the pregnant, which nature evacuates after birth.[20]

Thus, the key problem is, in Galen's view, the retention of the purges which are meant to occur following parturition. This is very dangerous, more dangerous than retaining the menses, because what is retained is not just excessive, but also nasty (not just *plêthos* but also *kakochymia*). During pregnancy, the embryo takes all the best humours for itself, so bad humours collect and require evacuation; which task, nature (*physis*), ably performs. Behind this is the basic Galenic (and Hippocratic) notion of the foetus being nourished by the surplus material which would otherwise be evacuated in menstruation; and also Galen's commitment to a provident nature, which is both innate in each individual and the transcendent creator of all things.[21] Though the particular juxtaposition of the two here creates a certain sympathy with Soranus of Ephesus' complaint about those physicians who, in arguing that menstruation is advantageous both to women's health and her reproductivity, refer to the beneficence of nature in ridding women of the excess they accumulate through sedentary domesticity by means of their menstrual purges.[22] As the Methodist points out: if nature is really provident, why does she not prevent the formation of female excess in the first place? The same question could be asked here: is it not contradictory to have a beneficent nature cast in a *remedial* role? Galen clearly does not think so – the explanatory flexibility of *physis* is one of her key virtues – and this basic understanding of the humoral economy in female reproduction is widely applied. It can, for instance, also be extended to deal with a case of pustules that break out while a woman in *Epidemics 2* is nursing, then disappear when she

19 Gal. *Hipp. Epid. 3* 3.77 (*CMG* V 10.2.1 166.20-25), cf. Hp. *Epid.* 3 17 (3 108.5-112.12 L).

20 Gal. *Hipp. Epid. 3* 3.77 (*CMG* V 10.2.1 167.6-12). On Galen and Sabinus see Smith, *Hippocratic Tradition*, 70-72.

21 On both these points of Galenic doctrine see Flemming, *Medicine*, esp. 293-328.

22 Soranus, *Gyn.* 1.27-8.

stops in the summer.[23] Galen suggests as a possible explanation for this, that, just as milk too uses up the material otherwise evacuated in menstruation, and just as the embryo uses up the best blood, so too the nursing infant, producing nasty excessive residues which are then excreted in pustular form.

Now Galen does have good Hippocratic foundations for this theory. Not only does it join up the Hippocratic model of foetal nutrition with the acknowledgement, made in both gynaecological and embryological works, that the failure of the lochia (the post-partum purges) is a certain cause of illness, and may well prove fatal; but there is a hint at his mechanism too in *Diseases of Women*.[24] Here the possibility is entertained that a pregnant woman may turn quite green, since she is constantly losing the pure (*akraiphnes*) part of her blood to her foetus, leaving her with insufficient to maintain her normal colour.[25] This dwindling of the blood also causes cravings for strange foods, morning sickness, and general debility. However, Galen clearly takes matters considerably further and switches attention from the symptoms – the greenness, nausea, cravings and so forth – to the systemic shift which underlies them. One important effect of the particular way he puts the Hippocratic pieces together is thus that pregnancy (and lactation) are no longer about the rectification of excess, but rather its exacerbation. It produces a problematic somatic situation which must be solved by something other than birth itself, it requires particular purgings (which nature takes care of).

The pathological potential of pregnancy is thus heightened, but Galen stops short of putting health and reproduction in direct opposition, as some Methodist physicians did (at least in theory). Soranus' *Gynaecology*, for instance, argues that pregnancy and childbearing are most unhealthy for the child-bearer, though most useful for the continuation of the species, and though he cites no specific precedents for this view that does not mean that he was the first to hold it.[26] In Galen's scheme, on the other hand, in the normal course of events – where pregnancy is followed by a straightforward birth and purgation – everything is safe. Moreover, he is willing to state elsewhere in the commentaries that pregnant women are less susceptible to disease than those who are not pregnant, though the effects of childbearing are deleterious in the longer-term as they weaken the woman's body and speed up the onset of aging (especially if pregnancy is frequent and not adequately spaced out).[27] Still, he has complicated the reproductive package. Birth in itself is not enough to achieve the requisite balance, proper purging must also occur, and he has to make recourse to his notion of a provident nature to attain this. Within the world of the *Epidemics*, in particular, by providing an explicit explanation for events as much as through the precise content of this explanation, Galen has, none the less, shifted the emphasis from a situation in which incidental and accidental risks of childbearing are simply recorded, to the institutionalisation of those risks,

23 Gal. *Hipp. Epid.* 2 2 (*CMG* V 10.1 223.32-224.30).

24 On the dangers of the non-appearance of the *lochia* see e.g. Hp. *Mul.* 1.35-40 (8 82.13-98.5 L) and *Nat. Puer.* 18 (7 502.14-18 L).

25 Hp. *Mul.* 1.34 (8 78.11-15 L).

26 Soranus, *Gyn.* 1.42, with one of the reasons for this opposition being that the foetus must take resources from the mother, which can only be damaging. It should be noted, however, that despite this theoretical position, the *Gynaecology* is basically a guidebook for successful reproduction.

27 Gal. *Hipp. Epid.* 2 2 (*CMG* V 10.1 222.23-5).

the embedding of them in the system. He has also, as hinted at earlier, introduced a problematic area into his own cosmological and physiological understandings, indeed there is a clear contradiction that comes out of this. For, in *On the Usefulness of the Parts*, Galen explicitly states that the systemic surplus of the female body actually produces the optimal conditions for foetal nourishment, one that in no way damages the woman.[28] Whereas now it is clear that she is, at least temporarily, harmed by pregnancy, indeed to describe what the foetus takes as surplus to her requirements is somewhat misleading. The strain of achieving Hippocratic and Galenic congruence is thus apparent on both sides.

There is a further aspect of this business which also requires attention. The Thasian woman by the water gives birth to a girl, and Hanson has noted the strong predominance of female offspring in cases where disease (and even death) follows either a living birth, or a miscarriage.[29] Indeed, the woman who lay by the Liars' Market in *Epidemics 3* is the sole example of death following a male birth.[30] It is, furthermore, being pregnant with a boy which is of most therapeutic effect, as both of the women cured in *Epidemics 2* show, though the wife of Gorgias bears (with considerable complications) a girl (and a superfoetation). Hanson explains this with reference to the Hippocratic association of the male with the right side, as being quicker to develop and move, so that pregnancy with a male is less taxing and the delivery is easier, since the child is stronger and better able to get out. Demand suggests a more practical explanation for this devastating sex-ratio: that post-partum care of both mother and child is of a lower quality if the child is a girl, on account of the lower value attached to her.[31] Of course, the two theories are not mutually contradictory. Indeed, it would undoubtedly be useful to have a physiological explanation for the results of neglect. However, while the lesser appreciation of a female addition to the family might well be thought to contribute to higher rates of infant mortality among girls, it is not at all clear why it should also affect the mother so badly (and it is the mother who is under discussion here).[32] Surely the reverse would be true: she needs to be kept healthy so that she can have a boy next time.

Galen certainly tends to agree with Hanson, but not entirely straightforwardly, and he has to be slightly creative in his interpretation of the *Epidemics* to do so. Though modern editions of *Epidemics 2* record a difficult birth, and incomplete purging, for the woman who gave birth to twin daughters, then swelled up and had nasty fluxes for several months; Galen's lemma does not.[33] He inserts dystocia into the sequence on the authority of 'some manuscripts' and the knowledge that the delivery of female babies 'takes longer' than that of males. With this addition the episode makes much more sense. The female twins are the root of the problems, their double femininity causing a long drawn out labour which keeps in the blood that should have been released, and exerts great pressure on the internal organs. Nor was the blood then purged, hence swelling resulted, along with various unpleasant flows

28 Gal. *UP* 14.6 (2. 299.10-301.25 Helmreich).

29 Hanson, 'Diseases', 48.

30 Hp. *Epid.* 3.1.12 (3.62.11-66.11 L).

31 Demand, *Birth*, 49.

32 For discussion of female infanticide and other issues affecting mortality see e.g. Cynthia Patterson, 'Not worth the rearing', *TAPA* 115 (1985), 103-22.

33 Hp. *Epid.* *2* 2.20 (5 92.8-12 L); cf. Gal. *Hipp. Epid.* *2* 2 (*CMG* V 10.1 230.4-11 for the lemma, and 230.12-234.7 for the discussion).

as equilibrium was gradually restored. Similar reasoning appears in relation to the household slave of Stymages who had problems with the mouth of her womb, her hip and leg, after giving birth to a daughter without any bleeding.[34] There is no mention of dystocia, but Galen brushes aside the symptom that is mentioned – the failure to bleed – and appeals directly to the 'longer *and* harder', labour that will have resulted from the sex of the child. This will have damaged the womb, and, by association, harmed the hip and leg too. Difficult birth also beset the woman by the Liars' Market, but there the implication is that the difficulties were the responsibility of the woman – who was only seventeen and giving birth for the first time – not the male child (who survived); as Galen simply remarks that she must have been pretty strong to have lasted as long as she did (she died on the fourteenth day after birth).[35] The dangers of dystocia are reinforced, however, whatever the cause.

In none of these cases does Galen offer a reason for the difficulties of female births, they are mobilised as explanatory, never explained. Nor is a comprehensive discussion of dystocia – its symptoms, causes, and remedies – to be found anywhere else in Galen's extant works. Moreover, what is to be found in his descriptions of normal birth rather precludes him from pursuing one of the main Hippocratic avenues to understanding why birth may not proceed easily, that is the foetal avenue. For, as Hanson stresses, it is the child who plays the major, active, role in the Hippocratic construal of parturitive mechanics, while the mother plays a relatively minor, passive part in proceedings.[36] The same cannot be said of Galen. In his physiological treatise, *On the Natural Faculties*, Galen provides an account of birth as, essentially, a function of the 'expulsive faculty' (*apokritikê dynamis*) of the uterus.[37] This faculty takes over from the womb's 'retentive faculty' (*kathektikê dynamis*) at the moment when the foetus has attained its proper size, and needs only a little assistance from the midwife (in dilating the mouth of the uterus), and rather more from the mother herself (in exerting the relevant non-uterine muscles). However, the baby itself makes absolutely no contribution to the process, even the moment of changeover from retention to expulsion is regulated by nature (with her providence and foresight), triggered by the completion of set tasks, not by that which has been completed.

Of course, it is the specific purpose and focus of *On the Natural Faculties* which shape this account of birth most decisively, that give it its facultative emphasis and rather abstract qualities. However, it is clearly the mother – both as a whole and a (uterine) part – who must take precedence in providing a Galenic explanation for *dystocia* in general, and its correlation with the sex of the child born in particular. The Hippocratic commentaries offer one such approach, in the form of a sexually differentiated feedback system that emerges as an overlay to the basic pathology of pregnancy already outlined.

In *On the Usefulness of the Parts* Galen constructs – relying on the Aristotelian assertion that the male is hotter than the female (his extra heat being coterminous with his superiority), and the Hippocratic aphorism which places male embryos mostly on the right and females mostly on the left – a complex account of embryonic sex determination in which two key

34 Gal. *Hipp Epid. 2* 4 (*CMG* V 10.1 343.36-345.19), cf. Hp. *Epid.* 2.4.5 (5 126.10-14 L).

35 Gal. *Hipp. Epid. 3* 2.15 (*CMG* V 10.2.1 105.5-107.11).

36 Ann Ellis Hanson, 'Continuity and change: three case studies in Hippocratic gynaecological therapy and theory', in S. B. Pomeroy (ed.), *Women's History and Ancient History* (Chapel Hill 1991), esp. 87-93.

37 Gal. *Nat. Fac.* 3.3 (*SM* 3 207.17-211.10).

factors are at work: the man's seed, and the woman's womb, with (again) the latter exerting the greater influence.[38] An asymmetric configuration of the blood-vessels descending to the generative parts in men and women makes both the right testicle (and the seed it pro-duces), and the right cavity of the uterus, hotter than their left-hand partners. So, hotter seed and a hotter receptacle both have a tendency to produce hotter, that is male, embryos; with the uterine cavity better able to effect this outcome as it has a longer period in which to achieve it. This crucial corner-stone of the Galenic system is much repeated and referred to, slightly amended and extended, in a whole range of his other works. One extension that is elaborated in the Hippocratic commentaries is that a woman's thermal situation can function as effect as well as cause, so establishing a kind of positive (or negative) feedback loop, in which the heat or cold of the child she has formed by the heat or cold of her womb may then have a more global effect on her somatic condition, with beneficial or harmful results as appropriate.

The clearest exposition of this feedback mechanism actually comes in Galen's commentary on the Hippocratic aphorism:

> γυνὴ ἔγευος ἢν μὲν ἄρρεν κύῃ, εὔχροός ἐστιν· ἢν δὲ θῆλυ, δύσχροος
>
> If a woman is pregnant with a male child, she has a good complexion; if with a female, a poor complexion.[39]

This observation is explained by reference to the colder quality of the female conception, a coldness which saps the pregnant woman's health and colour, in comparison to the healthier effects of the male.[40] The same understanding is reflected in Galen's commentary on the anonymous woman in *Epidemics 2* who had hip pains prior to conception which vanished during pregnancy only to return after she gave birth to a boy.[41] The discomfort, Galen suggests, was caused by excess cold, a situation alleviated both by the pregnancy itself – as that entails the retention of hot blood normally lost through menstruation – and, especially, by carrying a warmer, male, foetus. Returning to the point of birth itself, though Galen does not make the connection explicit himself, it could be that carrying a female foetus has detrimental consequences for delivery itself – as the cold diminishes the powers of both womb and woman, making the process longer if not harder.

Of course, that such an argument never actually surfaces in Galen's extant works, and the rather suspiciously Hippocratic turn of phrase in various crucial places, suggests that Galen did, implicitly, rely on more directly foetal causes of dystocia, as much as maternal ones. Certainly he expresses broad agreement with the cryptic comments in the *Epidemics* which are construed as referring to the superiority of the male foetus in terms of the speed of its development – of the basic formation of its limbs and articulation of its joints – its greater

38 Gal. *UP* 14.7 (2 302.1-310.7 Helmreich), and see also his account at *Sem.* 5.7-40 (*CMG* V 3.1 180.19-186.26). It should also be noted that, despite its absence from this account, Galen was committed to the existence and reproductive contribution of female seed.

39 Hp. *Aph.* 5.42 (4 546.4-5 L).

40 Gal. *Hipp. Aph.* 5.42 (17B 834-5 K).

41 Gal. *Hipp. Epid.* 2 2 (*CMG* V 10.1 227.36-228.12).

solidity and strength, and quicker movement.[42] Galen gives most attention to a passage in *Epidemics 6* which he comments on sentence by sentence, each one slightly at variance with the formulations transmitted to us, but which appears in Smith's most recent edition of the Hippocratic book as follows:

> ὅτι ἐν θερμοτέρῳ τὸ ἐν τοῖσι δεξιοῖσι, καὶ μελανθὲς διὰ τοῦτο, καὶ ἔξω αἱ φλέβες μᾶλλον. ξυνεκρίθη, ξυνέστη ὀξύτερον, κινηθέν, ἐμωλύνθε, καὶ βραδύτερον αὔξεται καὶ ἐπὶ πλείω χρόνον. ὅτι ἐστερεώθε καὶ χολωδέστερόν τε καὶ ἐναιμότερον, ᾗ τοῦτο θερμότερόν ἐστι τὸ χωρίον τῶν ζώων.

> Because what is on the right is in a hotter place it is darker because of that, and its blood vessels are more external. It collects, is constructed more quickly, moves, softens, and grows more slowly and for a longer time. Because it is solidified, it is more full of both bile and blood, so that it is the warmer area in animals.[43]

The attraction of the first sentence, which for once (in the Hippocratic writings) makes the heat of the right side explicit, is obvious, and Galen enjoys himself in elaborating on it with reference to *Aphorisms*, many of his own works, and both Parmenides and Empedocles.[44] The second sentence, other versions of which are also repeated elsewhere in the *Epidemics*, is rather more problematic.[45] A minor point of dissent appears as Galen delves deep into the rich exegetical tradition to note that Rufus claims that Diogenes of Apollonia challenged the consensus that the male foetus *moved* as well as being formed first; though Galen himself has not seen the work. But this is not the main issue. That is how to reconcile the two, apparently contradictory, parts of this compressed statement: how can what is formed more quickly actually grow more slowly, and for longer (especially if it is male)? His interpretative strategy, following many of his predecessors, is to separate the second half of the equation from the rest and have it refer to what happens after birth. Girls do reach sexual maturity faster than boys, who continue growing after they have stopped. For girls, however, there is no virtue in this developmental rapidity as there was for the hotter, quicker male foetuses in the womb. It is, instead, a function, or marker, of female inferiority, caused by their softness, wetness, and weakness.[46] The soft wet qualities of the female body even at maturity can be achieved much quicker than the hard, dry male body, which is just as well because the weakness of the natural faculty in women, which governs growth, means that this process grinds to a halt while the better males are still going strong. The commentary on the third sentence returns more to the themes of the first.[47]

42 This construal is made in light of the rather clearer exposition of Hippocratic embryology in *On the Nature of the Child*, which includes (*Nat. Puer.* 18; 7 498.27-506.2 L) the statement that the male foetus is formed in 30 days (maximum) and the female in 42 (maximum).

43 Hp. *Epid.* 6.2.25 (234.10-15 Loeb Hippocrates vol. VII, cf. 5 290.7-12 L); cf. also Gal. *Hipp. Epid. 6* 2.46 (*CMG* V 10.2.2 118.20-22).

44 Gal. *Hipp. Epid. 6* 2.46 (*CMG* V 10.2.2 118.23-121.11).

45 See Hp. *Epid.* 6.8.6 (5 344.13-15 L) for repetition, though Galen transports the sentence to Book 2 3.17 (*CMG* V 10.1 297.25-7). On this line itself see *CMG* V 10.2.2 121.15-123.6.

46 Compare, however, the different emphases given to these factors at 2.13.7 (*CMG* V 10.1 297.28-298.39) and 6.2.46 (*CMG* V 10.2.2 121.15-123.6).

47 *CMG* V 10.2.2 123.9-18.

Without the kind of understanding of birth outlined in the Hippocratic treatise *On the Nature of the Child*, in which the foetus, now unable to get the nourishment it needs from the mother, breaks the membranes which surround it then fights its way out of the womb entirely unaided, these discussions of its strength and activity are of little consequence.[48] Despite this, it seems likely that, when Galen simply states that the delivery of a girl is a longer, more arduous, business than that of a boy, his case mostly rests on the inferior qualities of the female foetus, on the intrinsic and pervasive hierarchy of humanity. In exegetical terms then, it appears that Galen may sometimes find the master's ideas more compelling than his own: which is, in a sense, to invert Smith's claim that Galen's Hippocratism owes more to himself than Hippocrates, and to see the Hippocratism that can overcome the Galenism. Of course the contrast lies, at least to some extent, in the angle of approach. Coming from the Hippocratic side – as Smith does – emphatically reveals Galen's manipulation of the tradition he inherited, whereas coming from the Galenic side reveals the Hippocratic anomalies lodged within it. There is also a difference between centre and periphery at work here. Many of the particular issues Galen confronts in the *Epidemics* in relation to pregnancy, miscarriage, birth and its aftermath, are frankly peripheral to his main concerns, and there is less baggage of his own to impede the assimilative pull of the texts he is engaged with. On points which are more crucial to his own system Galen is much more unbending, reversing the direction of assimilation. So, for example, he makes constant, unwarranted, use of his own notion of a provident nature in the commentaries, supported only by a couple of rather desperate attempts to pressgang Hippocrates into belief in nature's bounty.[49]

Nor is it just a Hippocratic pull that is at issue here, especially in relation to the female case histories. There is a societal pull too; a sense in which the Hippocratic version of the events around birth fits better with prevalent classical presumptions about the subordinate and passive qualities of women than the Galenic story would, if told in full. The implications of Galen's more womb- (if not woman-) centred theories are no match for the sheer obviousness of the fact that giving birth to girls is going to be harder work. Everything female is always worse, by definition. The point is underlined by the acrobatics Galen has to perform in order to keep both female foetus and infant from ever challenging their male counterparts for superiority in the growth department, despite so clearly outperforming them. He may explicitly disdain some contemporary values, but the hierarchy of the sexes is as certain for Galen as anyone else.

The main point that emerges from both lines of enquiry pursued here – into Galen as Hippocratic exegete and reluctant obstetrician – is, therefore, the complexity and flexibility of Galen's approach. His main concern may always be to secure his own authority, and his Hippocratism was a key means to that end, but it was not the only means, nor was its subordination to his own purposes either as straightforward as Smith sometimes suggests, or particularly distinctive in the world of classical medicine. The Hippocratic Corpus had been made malleable long before Galen: if not at the moment when such a heterogeneous body of material was brought together, then as successive layers of interpretation and manipulation accumulated around it. Even if Galen had wanted to comment on the Corpus in its own terms,

48 Hp. *Nat. Puer.* 30 (7 530.20-538.28 L).

49 See e.g *Hipp. Epid. 6* 6.5.1 (*CMG* V 10.2.2 253.16-254.16).

it is not clear that he could have, or that it was still possible (if it ever had been) to isolate the Hippocratic writing itself from all the ways it had been understood and built on, debated and deployed, since. He did want to display his mastery of this tradition, in its entirety, to highlight a pattern of convergence between his own and Hippocrates' ideas and prescriptions, a pattern that contrasted with the fabric of classical medical discourse more generally, and in which the priority of discovery, on the Hippocratic side, was matched (if not superseded) by his ability to explain the discoveries, to make links, to fit them into larger explicatory webs. This convergence could, moreover, be eased by selective compromises on either side, and the explanatory mechanisms could also be drawn from a range of sources. The most important thing was the multiplicity and thickness of the connections made, the ways in which points could be joined up and made sense of, not absolute purity or consistency. Some items, understandings or principles, were fixed, both within 'the Hippocratic' and 'the Galenic' parts of the equation, but around them there was plenty of (though not unlimited) room for manoeuvre, as the pathology of pregnancy demonstrates.

King's College London.

THE USES OF GALEN IN ARABIC LITERATURE

GOTTHARD STROHMAIER

At the turn of the first millenium of the Christian era the great Muslim scholar al-Bīrūnī (973-1048) was fortunate enough in having the opportunity to explore the customs, the sciences and the religion of the Indians. He found all this bizarre and, though being himself a native of Choresmia in Central Asia, he surely would have agreed with Rudyard Kipling's famous verse: 'Oh, East is East, and West is West, and never the twain shall meet.' Nevertheless in his huge monograph on India[1] he tried to explain the Hindus' idolatry, their polytheism and their belief in metempsychosis by comparing them with what he knew from Greek literature about pre-Christian paganism in the Ancient World. He drew upon what had been translated in Bagdad in the ninth century AD, namely Plato, Aristotle, Aratus, Ptolemy, and last but not least Galen;[2] and it is astonishing to see how he found the scattered information that interested him out of the bulky corpus of the Pergamene physician. They are taken from the *Protrepticus,*[3] *De compositione medicamentorum secundum locos,*[4] *De compositione medicamentorum per genera,*[5] *De moribus,*[6] *De demonstratione,*[7] from *That the first mover is unmoved (In primum movens immotum)*',[8] and the commentary on the Hippocratic Oath.[9] It was from Galen that al-Bīrūnī could learn something of Greek mythology, about Kronos, who alone was eternal among the gods, and his son Zeus, whom Kronos was about to devour, or about the statues of Hermes and Asclepius.

The German orientalist Eduard Sachau who published an edition and an annotated translation of al-Bīrūnī's work in 1887 and 1888 in London was only partly successful in identifying them in Kühn's edition. One sees how important it may be for an arabist to have

1 *Fī taḥqīq mā li-'l-Hind,* ed. E. Sachau (London 1887); *Alberuni's India,* trans. E. Sachau, 2 vols (London 1888).

2 G. Strohmaier, 'Der syrische und der arabische Galen', *ANRW* II, 37.2 (1994) 2007-11.

3 *Exhortatio ad medicinam (Protrepticus)* 9,2: ed. A. Barigazzi, *CMG* V 1,1 (Berlin 1991) 132,16-21; cf. *India,* 16,18-17,1 (trans., vol. 1, 34-35); cf. F. Rosenthal, 'An ancient Commentary on the Hippocratic Oath', *BHM* 30 (1956) 62.

4 Galen, *De comp. med. sec. loc.* IX 4: XIII 268, 271 K. cf. *India,* 46,15-47,1 (trans., vol. 1, 95-96).

5 Galen, *De comp. med. sec. gen.* VII,9; XIII 995-96 K. cf. *India* 61,12-15; 72,21-73,2 (trans., vol. 1, 127, 151).

6 *Kitāb al-akhlāq li-Djālīnūs,* ed. Paul Kraus, in *Madjallat kullīyat al-ādāb bi-'l-djāmi'at al-miṣrīya* 5 (1937) 40; cf. *India* 59,20-60,5 (trans., vol. 1, 123-24).

7 *India,* 47,15-16 (trans., vol. 1, 97).

8 *India,* 164,1-2 (trans., vol. 1, 320); cf. G. Fichtner, *Corpus Galenicum. Verzeichnis der galenischen und pseudo-galenischen Schriften* (Tübingen 1995) no. 397.

9 *India,* 17,1. 16. 18; 109,10-11; 283,10-13 (trans., vol. 1, 35-36, 222; vol. 2, 168); cf. Rosenthal, Commentary, above, note 3, 61-63.

as easy an access to Galen as, say, to Plato or Aristotle. Al-Bīrūnī was not a physician but Galen was present in the minds of the whole erudite class of Muslim society as a natural scientist of paradigmatic stature. Thus we may expect to find quotations from his works anywhere in the vast scholarly Arabic literature of the Middle Ages. I cannot claim to be exhaustive in this paper in any way; rather I shall present some typical examples in order to show what might still be expected and also the pitfalls to be encountered.

At present there seems to be not much hope of finding much in the way of new entire Galenic works in Arabic translation.[10] But there remains the urgent task of collecting the fragments. Many examples of this kind of tradition were already collected by Moritz Steinschneider,[11] Manfred Ullmann,[12] Fuat Sezgin,[13] and, for the philosophical writings, more recently by Mauro Zonta,[14] who has also included Hebrew testimonies hitherto largely overlooked.

Iwan von Müller in his attempt to reconstruct the lost work *On Proof (De demonstratione)* had already incorporated some material from the Latin Averroes, the Latin Moses Maimonides and the Latin *Liber continens* of Rhazes.[15] This may now be enlarged considerably from the Arabic sources themselves. Many new fragments are contained in Rhazes' (d. 925 or 935) *Doubts on Galen* (*Shukūk ʿalā Djālīnūs*), an outstanding example of the critical and self-confident position of this great Persian physician. Rhazes deals at the beginning at length with Galen's *On Proof (De demonstratione)*, in his opinion the most useful book after the revealed scriptures. From the other quotations which can be identified in other existing works we are able to see that Rhazes gives them sometimes literally and sometimes according to the sense only.[16] The same holds true for the *Kitāb al-ḥāwī*, the 'Liber continens' of the Latin tradition, where Ursula Weisser has checked the quotations from the *Method of Healing (Methodus medendi)*.[17]

10 G. Strohmaier, Der syrische und der arabische Galen, above, note 2, 2016.

11 M. Steinschneider, 'Die griechischen Ärzte in arabischen Übersetzungen', *Archiv für pathologische Anatomie und Physiologie und für klinische Medizin* 124 (1891) 268-468, 487 (reprint in: *Islamic Medicine*, vol. 15: *Hippocrates in the Arabic Tradition. Texts and Studies* (Frankfurt am Main 1996), Publications of the Institute for the History of Arabic-Islamic Science, 116-58, 177).

12 M. Ullmann, *Die Medizin im Islam* (Leiden, Cologne 1970), Handbuch der Orientalistik. Erste Abt., Erg.-Bd. VI, erster Abschnitt, 38-65, 344-45; additions in: idem, *Die Natur- und Geheimwissenschaften im Islam* (Leiden, Cologne 1972), Handbuch der Orientalistik. Erste Abt., Erg.-Bd. VI, zweiter Abschnitt, 455-56.

13 F. Sezgin, *Geschichte des arabischen Schrifttums*, vol. 3 (Leiden 1970) 78-140; additions, ibid., vol. 5 (Leiden 1974) 404-16; vol. 7 (Leiden 1979) 375-91.

14 M. Zonta, *Un interprete ebreo della filosofia di Galeno. Gli scritti filosofici di Galeno nell'opera di Shem Tob ibn Falaquera* (Turin 1995) 4-20.

15 Iwan von Müller, 'Über Galens Werk vom wissenschaftlichen Beweis', *ABAW* 20, 2 (1895) 405-78.

16 G. Strohmaier, 'Bekannte und unbekannte Zitate in den *Zweifeln an Galen* des Rhazes', *Text and Tradition. Studies in Ancient Medicine and its Transmission Presented to Jutta Kollesch,* ed. K.-D. Fischer, D. Nickel and P. Potter (Leiden, Boston, Cologne 1998) 263-87; for other quotations from *De demonstratione* and from *That the first mover is unmoved* in the alchemical Djābir corpus see Paul Kraus, 'Jābir Ibn Ḥayyān. Contribution à l'histoire des idées scientifiques dans l'Islam'. Vol. 2: Jābir et la science grecque, *Mémoires présentées à l'Institut d'Égypte 45* (1943) 328-29 (reprint Hildesheim, Zurich, New York 1989); cf. F. Sezgin, *Geschichte* (above, note 13), vol. 5, 129.

17 U. Weisser, 'Zur Rezeption der Methodus medendi im Continens des Rhazes', in *Galen's Method of Healing. Proceedings of the 1982 Galen Symposium*, ed. F. Kudlien and R. J. Durling (Leiden, New York, Copenhagen, Cologne 1991) 123-46; eadem, 'Die Zitate aus Galens De methodo medendi im *Ḥāwī* des Rāzī', in *The ancient tradition in Christian and Islamic Hellenism. Studies on the Transmission of Greek Philosophy and Sciences dedicated to H.J. Drossaart Lulofs on his ninetieth birthday*, ed. G. Endress and R. Kruk (Leiden 1997) 279-318.

Sometimes a defective Greek textual tradition may at least partly be completed by quotations found in Arabic authors. This was possible in Phillip De Lacy's edition of *On the opinions of Hippocrates and Plato (De placitis Hippocratis et Platonis)*, where an extract by Rhazes may be compared with the more complete version of Ibn al-Muṭrān (d. 1191).[18] Another example has now appeared in Vivian Nutton's recent edition of *On my own opinions (De propriis placitis)* where the defective Greek text could be completed by pieces from the Hebrew and Latin tradition, both based on a lost Arabic translation.[19]

In a passage of the *Chronology* missing in the Sachau edition,[20] but discovered later in two more complete manuscripts, we find a description of St. Elmo's fire, an electrical phenomenon to be seen on the high seas, which according to popular belief was held to announce salvation from imminent shipwreck. Al-Bīrūnī adds that this is confirmed by eye-witness accounts and also by Galen.[21] In a little German anthology I put together from the works of al-Bīrūnī I was still unable to verify this account,[22] but now with Vivian Nutton's edition of *On my own opinions* it became clear that this is an allusion to this very tract or perhaps to another where he had dealt with this personal experience at sea more circumstantially.[23]

In his huge astronomical handbook *Al-Qānūn al-Masʿūdī* al-Bīrūnī mentions a talk Galen had with his father about the heliacal risings of certain stars when his father praised him for his knowledge of astronomical matters. Al-Bīrūnī adds that Galen is held in high esteem also nowadays although some people do criticize him.[24] He does not indicate the work from where he had taken this story. But now with the forthcoming edition of Galen's commentary on Hippocrates' *Airs, Waters, and Places* we are in a position to judge that al-Bīrūnī was correct. The question was why Sirius appears in Pergamum two days later than in Alexandria, a phenomenon that most astronomers in Asia were unable to explain.[25]

In a passage in his *Exhaustive Treatise on Shadows* al-Bīrūnī mentions that Galen rightly criticized Hippocrates for his explanations of the allegedly hereditary character of skull deformations among the Scythian population. This is again taken from Galen's commentary on *Airs, Waters, and Places*. But here it is obvious that al-Bīrūnī is quoting from memory.

18 *Galeni De placitis Hippocratis et Platonis*, *CMG* V 4,1,2, ed. P. De Lacy, 3 vols (Berlin 1978-84) vol. 1, 42-46 and 70-77.

19 *Galeni De propriis placitis*, ed. V. Nutton, *CMG* V 3,2 (Berlin 1999) 31-33 and 56-61.

20 *Chronologie orientalischer Völker von Albêrûnî*, ed. E. Sachau, Leipzig 1923.

21 A. B. Khalidov, 'Dopolneniya k tekstu "Khronologii" al-Bīrūnī po leningradskoy i stambulskoy rukopisyam', *Palestinskiy sbornik* 67 (1959) fasc. 4, 160-61.

22 Al-Bīrūnī, *In den Gärten der Wissenschaft. Ausgewählte Texte aus den Werken des muslimischen Universal-gelehrten*, trad. G. Strohmaier, 2nd edn (Leipzig 1991) 113.

23 *Galeni De propriis placitis, CMG* V 3,2, p. 58: and comm. 139-40; cf. M. Zonta, *Un interprete ebreo*, above, note 14, 103-08.

24 Al-Bīrūnī, Al-Qānūn al-Masʿūdī (Hyderabad 1954-56) 1139; Russian translation in: *Abu Raykhan Beruni, Izbrannye proizvedeniya,* vol. 5.2 (Tashkent 1976) 331.

25 *Galeni In Hippocratis De aere aquis locis commentariorum versio Arabica*, *CMG* Suppl. Or. V, ed. G. Strohmaier (Berlin in press); Ms. Cairo, *Ṭalʿat, ṭibb* 550, fol. 65r20-65v2 and 65v11-66r15.

He speaks of 'the broad-headed' instead of 'the long-headed', perhaps being influenced by a local Choresmian custom of flattening the heads of the new born.[26]

The risk that the quotations are not exact when reproduced from memory is especially virulent when the content has an anecdotal character. The following examples can all be checked against the preserved original version.

The physician Abū Saʿīd ibn Bukhtīshūʿ (d. after 1058) in a tract *On the Healing of the Maladies of Soul and Body* has many references to Galen. He was particularly impressed by the story of how Galen diagnosed the lovesickness of a Roman lady from her pulse, giving correctly the current Arabic title *Fī nawādir taqdimati l-maʿrifa (Prognostical Anecdotes*) that the translators had given to *On prognosis (De praecognitione)*. But the narrative is completely distorted according to the incentives of hagiography. Galen here helps a young girl who has fallen in love with a youth who used to ride on horseback in her father's entourage. Galen conferred with the mother and arranged the wedding, whereupon the grateful young couple revered him like a father.[27] Such a happy ending is missing in the original story, which was without doubt closer to reality, for Galen himself in *On prognosis* says nothing about a successful therapy. But we understand here how legends may originate from a kernel of historical truth.[28]

Niẓāmī ʿArūḍī, a Persian story-teller of the twelfth century, collected in his *Čahār Maqāla* anecdotes of famous personalities, including Galen. He recalled how he had healed the sophist Pausanias from Syria who had fallen out of his carriage, which led to paralysis and numbness of several fingers. Galen tells the story often and ascribes his success to his anatomical knowledge.[29] In Niẓāmī the patient is one of the notables of Alexandria, but Galen's diagnosis and therapy are given quite correctly.[30]

A special chapter of Galen's heritage in Arabic literature is opened with the so-called gnomologia. They consist of a mixture of wise sayings and anecdotes ascribed to great personalities, as a rule without any connection with their original works.[31] The name of Galen is missing in the Greek collections, so far as I know, as he was, obviously, not famous enough here to be included. But he is present in at least two Arabic gnomologia, which is a sign of Galen's popularity in the Islamic society, namely in Ḥunayn's *Nawādir al-falāsifa* (*Anecdotes*

26 *Iqrār al-maqāl fī amri ẓ-ẓilāl*. in *Rasāʾil al-Bīrūnī* (Hyderabad 1948) 38,15-18; Al-Bīrūnī, *The exhaustive Treatise on Shadows*, trans. and comm. E. S. Kennedy, 2 vols (Aleppo 1976) vol. 1, 77-78; vol. 2, 29-30; cf. *Galeni In Hippocratis De aere*, above, note 25; Ms. Cairo, *Ṭalʿat, ṭibb* 550, fol. 80r20-81r7.

27 Abū Saʿīd ibn Bukhtīshūʿ, *Risāla fī ṭ-ṭibb wa-l-aḥdāthi n-nafsānīya. Über die Heilung der Krankheiten der Seele und des Körpers*, ed. F. Klein-Franke (Beirut 1977) Arabic 59-60 (trans. 94).

28 Cf. *Galeni De praecognitione 6*: *CMG* V 8,1, 100-03 and the commentary *ad loc.*.

29 Cf. *De anatomicis administrationibus* III,1; XV,4: *Anatomicarum Administrationum Libri qui supersunt novem. Earundem interpretatio arabica Hunaino Isaaci filio ascripta*, ed. I. Garofalo, 2 vols, vol. 1 (Naples 1986) 133-36 = II 343-45 K.: *Sieben Bücher Anatomie des Galen*, ed. M. Simon (Leipzig 1906) 2 vols, vol. 1, 307 (German transl., vol. 2, 222-23); *De locis affectis* I,6; III,2.14: VIII 56-61; 138; 213-214 K.; *De optimo medico cognoscendo* 9,10-11: ed. A. Z. Iskandar, *CMG* Suppl. Or. IV (Berlin 1988) 106-09.

30 Trans. E. G. Browne (London 1921) no. 41, p. 95; cf. A. M. Zafari, *Nezami Aruzi. Tschahar Maqaleh. Psychosomatische Aspekte in der mittelalterlichen Medizin Persiens*, Arbeiten der Forschungsstelle des Instituts für Geschichte der Medizin der Universität zu Köln 53 (Cologne 1990) 44-45.

31 G. Strohmaier, 'Das Gnomologium als Forschungsaufgabe', in *Dissertatiunculae criticae. Festschrift für Günther Christian Hansen*, ed. Ch.-F. Collatz, J. Dummer, J. Kollesch and M.-L. Werlitz (Würzburg 1998) 461-71.

of Philosophers)[32] and in al-Mubashshir's *Mukhtār al-ḥikam wa-maḥāsin al-kalim* (*The Choicest Maxims and Best Sayings*).[33]

As usual in this kind of literature the authors are quoted without any necessary basis in their genuine works. But in one instance we can locate a passage out of which an anecdote in the literary form of the so-called *chreia* has been formed. In the *Protrepticus* we read of the athlete Milo of Croton who carried a slaughtered bull across the stadium, upon which Galen remarks that the soul of the bull was carrying it when it was alive and with much more speed. The *chreia* gives this a more narrative form:

> He saw a man honoured by the kings for the strength of his body. He asked about the great thing he had done, and they answered: 'He carried a slaughtered bull from the middle of the palace bringing it outdoors', whereupon he said to them: 'And the soul of the bull did carry it without showing any virtue in having it carried.'[34]

On the whole this gnomological material is to be viewed with the greatest suspicion, for the constant reshuffling of these little pieces in the textual tradition has led to the utmost confusion in the names of the authorities to whom the wise sayings are ascribed. In the field of historiography, however, we stand on firmer ground. It was characteristic of Arabic historiography to reproduce various and often contradictory materials from different sources without bothering about harmonizing them.

The medical historian Ibn abī Uṣaybiʿa quotes at length from Galen's *On Examinations by which the best Physicians are recognized (De optimo medico cognoscendo*). This work is lost in Greek, but one could assume that the author was copying from a Galenic work that lay before him, and Max Meyerhof was right in including these passages in his article 'Autobiographische Bruchstücke Galens aus arabischen Quellen.'[35] Now the edition by Albert Z. Iskandar in the *Supplementum Orientale* of the *Corpus Medicorum Graecorum* has shown that the passages in Ibn abī Uṣaybiʿa are reproduced so faithfully that they can be used as textual sources alongside the manuscripts.[36]

This gives us confidence that another passage he takes from another lost tract *On the Avoidance of Grief (De indolentia*) is genuine as well. Galen here tells us that he had lost manuscripts of Aristotle, Anaxagoras and Andromachus in a big fire, obviously that of the Temple of Peace in 192 AD.[37] Ibn abī Uṣaybiʿa and other historians have another story, of an incident when on the streets of Rome a quack from Aleppo pretended to heal toothache

32 K. Merkle, *Die Sittensprüche der Philosophen 'Kitâb âdâb al-falâsifa' von Ḥonein ibn Isḥâq in der Überarbeitung des Muḥammed ibn ʿAlī al-Anṣârî*, Phil. Diss. (Munich, Leipzig 1921) 25 and 61; the same collection in: Ibn abī Uṣaybiʿa, *ʿUyūn al-anbāʾ fī ṭabaqāti ʾl-aṭibbāʾ*, ed. August Müller (Cairo 1882) vol. 1, 88-89.

33 Ed. ʿA. Badawi (Madrid 1958) 293-96; the same collection in: Ibn abī Uṣaybiʿa, *ʿUyūn al-anbāʾ fī ṭabaqāti ʾl-aṭibbāʾ*, vol. 1, 89-90.

34 Al-Mubashshir, 296,6-8; *Exhortatio ad medicinam (Protrepticus)* 13,3, *CMG* V 1,1, ed. A. Barigazzi (Berlin 1991) 144-47; for another parallel cf. al-Mubashshir, 294,8-12, and *Exhortatio* 6,1.2: *CMG* V 1,1, 120,31-122,11.

35 M. Meyerhof, 'Autobiographische Bruchstücke Galens aus arabischen Quellen', *AGM* 22 (1929) 72-86 (reprint in: *Islamic Medicine, vol. 20: Galen in the Arabic Tradition.* Texts and Studies III (Frankfurt am Main 1996) 208-22).

36 Galen, *De optimo medico cognoscendo*, *CMG* Suppl. Or. IV, 15-16.

37 Ibn abī Uṣaybiʿa, *ʿUyūn al-anbāʾ fī ṭabaqāti ʾl-aṭibbāʾ*, vol. 1, 84,31-85,2.

by pulling worms out of the patients' teeth. This was taken from *On Diseases that are difficult to heal*, which is lost in both Greek and Arabic.[38]

Another interesting piece is to be found in Ibn al-Qifṭī. He reports on the authority of Galen a custom in Nubia, where venesection was to be carried out on men by women and on women by men. Galen had studied in Alexandria and it may be that he himself had travelled up the Nile or that a fellow student hailing from this region had told him this.[39] Ibn al-Qifṭī quotes from additions Ḥunayn had made to an introduction to a Galenic tract on venesection. He does not indicate the exact title but says only that it was not the famous little tract on venesection but a bigger one. Perhaps Ḥunayn thought it useful to insert this interesting detail from some other work of Galen unknown to us.

Galen was also partly responsible for the Platonic traditon in Islam through his summaries of Plato's dialogues. Ḥunayn had translated them while we have no trace whatsoever of them in the Greek tradition.[40] I say partly, for full translations of Plato were also available.[41] But the rôle Galen played in the transmission of Plato's philosophy to the Muslims should not be underestimated. As Dimitri Gutas has proved, al-Fārābī, the great champion of Peripatetic philosophy in the tenth century, made use of Galen's summary of Plato's *Laws*.[42]

While the summary of the *Timaeus* is preserved fully in Arabic translation,[43] only rare fragments remain of the other summaries. Among them is a famous statement about the Christians, who were emerging more and more from the underground in Galen's time and thus not escaping his attention. He says that they lead the life of philosophers while believing in fables and myths. This was very interesting for Muslims and Oriental Christians alike who quoted this on various occasions.[44] There is a curious difference. Some authors, mainly Christians, attribute this to the summary of the *Phaedo*, while others, mainly Muslims, locate this in Galen's summary of the *Republic*.[45] This does not exclude the possibility that Galen might have repeated his statements in his summary both of the *Phaedo* and of the *Republic*.

Averroes (1126-1198) in his commentary on the *Republic*, which is preserved in a Hebrew translation, quarrels with Galen over his casual comments on the text, e.g. when he stated that an average number of one thousand guardian warriors would not suffice for the empire of his

38 Ibn abī Uṣaybiʿa, *ʿUyūn al-anbāʾ fī ṭabaqāti ʾl-aṭibbāʾ*, vol. 1, pp. 81-82; translation in: G. Strohmaier, 'Der Arzt in der römischen Gesellschaft. Neues aus der arabischen Galenüberlieferung', *Acta Conventus XI 'Eirene', Warsaw 1971*, 70-71 (reprint in: idem, *Von Demokrit bis Dante. Die Bewahrung antiken Erbes in der arabischen Kultur* (Hildesheim, Zurich, New York 1996) 84-85; almost identical in: Ibn Ǧulǧul, *Les générations des médecins et des sages*, ed. F. Sayyid (Cairo 1955) 43; and Ibn al-Qifṭī, *Taʾrīkh al-ḥukamāʾ*, ed. J. Lippert (Leipzig 1903) 124.

39 *Taʾrīkh al-ḥukamāʾ*, ed. J. Lippert, 132.

40 G. Bergsträsser, 'Ḥunain ibn Isḥāq Über die syrischen und arabischen Galen-Übersetzungen', *Abhandlungen für die Kunde des Morgenlandes* 17.2 (1925), no. 124.

41 Cf. Abū Saʿīd ibn Bukhtīshūʿ, *Risāla fī ʾṭ-ṭibb wa-ʾl-aḥdāthi ʾn-nafsānīya*, ed. F. Klein-Franke, above, note 27, 46-47 (trans., 79-80).

42 D. Gutas, 'Galen's Synopsis of Plato's Laws and Fārābī's *Talḫīṣ*.', in Endress and Kruk, *The ancient tradition*, above, note 17, 101-19.

43 *Galeni Compendium Timaei Platonis aliorumque dialogorum synopsis quae extant fragmenta*, ed. P. Kraus and R. Walzer (London 1951).

44 R. Walzer, *Galen on Jews and Christians* (Oxford 1949) 15-16, 57.

45 S. Gero, 'Galen on the Christians. A Reappraisal of the Arabic Evidence', *Orientalia Christiana Periodica* 56 (1990) 371-411.

time or when he failed to understand why the weddings of the guardians should be repeated when they do not result in an offspring. Averroes also mentions in this context that Galen had written a book about the drawing of lots, but the Hebrew version is not altogether clear.[46]

In two instances the Arabic translation has helped to evaluate two pieces of the Greek textual tradition, one printed in Kühn under the title *How to convict Malingerers (Quomodo morbum simulantes sunt deprehendendi)* and *On the Substance of the natural Faculties (De substantia facultatum naturalium)*. They are not mentioned by Ḥunayn or by Galen himself in his bibliographical writings. But this is hardly surprising because these small tracts were excerpted in Byzantine times from larger works, namely from the commentaries on the *Epidemics* and from *On my own opinions*.[47] There may be other little pieces in Greek manuscripts purporting to be Galenic which could be rehabilitated by trustworthy Arabic authors who quote from them or allude to their content under the title of a work otherwise lost but known to be genuine.

The Christian translator Qusṭā ibn Lūqā in a correspondence with his Muslim Maecenas ʿAlī ibn Yaḥyā al-Munadjdjim about religious matters mentions that Galen had adduced many arguments in favour of the validity of dreams in his commentary on Hippocrates' *On humours*, but unfortunately without giving more details.[48] There is a doxographical chapter in an anonymous Bagdad manuscript where in the name of Galen various types of dreams are connected with the four humours.[49] The editor Helmut Gätje reminded us of a very little tract *On Diagnosis from Dreams (De dignotione ex insomniis)* contained in Kühn's edition but not identical with this Arabic text.[50] We must leave open at least the possibility that *On Diagnosis from Dreams* is nothing other than a part of the commentary on Hippocrates' *On Humours*. Perhaps we will see more when the text of a manuscript in Mashhad (Iran) described by Fuat Sezgin as this very commentary is examined more closely.[51]

But one must be careful. Some lines before Qusṭā ibn Lūqā in the same letter to ʿAlī ibn Yaḥyā reproduced from *On Examinations by which the best Physicians are recognized* Galen's account of how he predicted the miscarriage of a woman and of how he was highly praised by everyone except the husband, who remained unimpressed.[52] But here one grave divergence appears which seems to cast a general doubt on Qusṭā ibn Lūqā's reliability. The husband is, according to the French translation by Paul Nwiya and Samir Khalil, a 'wine merchant', while Galen characterizes him simply as 'beastly' (*bāhimīyan*). But the difference is easily removed. The corresponding Arabic word in Qusṭā ibn Lūqā should be punctuated not as *ḫammāran* ('a wine merchant') but rather as *ḥimāran* ('an ass'). We must also keep

46 *Averroes' Commentary on Plato's Republic*, ed. E. I. J. Rosenthal (Cambridge 1969) 46,6-12; 55,23-24; 56,24-26; 105,1-3 (trans. 152; 168; 170; 249).

47 Cf. Galens 'Simulantenschrift', ed. K. Deichgräber and F. Kudlien, in *Galens Kommentare zu den Epidemien des Hippokrates. Indizes der aus dem Arabischen übersetzten Namen und Wörter*, *CMG* V 10,2,4, ed. F. Pfaff (Berlin 1960) 107-20; *De propriis placitis*, *CMG* V 3,2,14.

48 *Une correspondance islamo-chrétienne entre Ibn al-Munağğim, Ḥunayn ibn Isḥāq et Qusṭā ibn Lūqā*, ed. Kh. Samir and P. Nwyia, Patrologia Orientalis 40.4, no. 185 (Turnhout 1981) 636-37.

49 H. Gätje, *Studien zur Überlieferung der aristotelischen Psychologie im Islam* (Heidelberg 1971) 91-92, 136-39.

50 VI 832-35 K.

51 F. Sezgin, *Geschichte*, above, note 13, vol. 3, 35 and 130.

52 *Une correspondance islamo-chrétienne*, above, note 48, 636-37; cf. *Galeni De optimo medico cognoscendo* 13, *CMG* Suppl. Or. IV, 130-33.

in mind here that Qusṭā ibn Lūqā was a very capable translator in his own right, and therefore was not dependent on the wordings of a translation from the Ḥunayn school. We should not therefore expect him necessarily to have the same expression *bāhimīyan* ('beastly') for the original Greek word.

These examples point a brief moral. Before passing any rash judgment about the value or reliability of information preserved in an Arabic author, the classicist would be wise always to consult an arabist first.

Berlin

GALEN'S LOST OPHTHALMOLOGY AND THE *SUMMARIA ALEXANDRINORUM*

EMILIE SAVAGE-SMITH

Nearly all major collections of Arabic manuscripts have a number of Galenic treatises that are stated to be *Jawāmi' al-Iskandarānīyīn* ('Summaries of the Alexandrians') or *jawāmi'* of Galenic treatises said to have been translated into Arabic by Ḥunayn ibn Isḥāq in the ninth century. Occasionally a compiler of the *jawāmi'* is named, such as Thābit ibn Qurrah (d. 901) or the enigmatic figure known in Arabic as Yaḥyá al-Naḥwī. In most instances, however, the compiler is anonymous and it is uncertain whether the summary was originally made in Greek, Syriac, or Arabic. The statement in a manuscript that a treatise was translated by Ḥunayn ibn Isḥāq is not sufficient evidence, by itself, for maintaining that it was translated from the Greek and that Ḥunayn made the translation, for virtually every manuscript copy of a work claiming a Greek origin has such a statement. The word *jawāmi'* can be here used as both a singular and a plural, while, occasionally, different manuscript copies will employ synonyms (*talkhīṣ*, *jumal*, *jumlah*, or *ikhtiṣār*).

Considerable scholarly attention has been drawn to those *jawāmi'* that are said to be summaries of the so-called 'Canon' of sixteen books read by the Alexandrians.[1] But the discussion should be expanded to include the entire group of Arabic summaries of Galenic treatises which apparently circulated in the ninth century, most of which the copyists also associated with the Alexandrians. A brief review of the fragility of the evidence for confidently associating any of them with the Alexandrians and a reminder of the testimony of Ḥunayn ibn Isḥāq as to the existence of Galenic summaries in his day, will be followed by a list of those *jawāmi'* that were recorded by Ḥunayn or are known to exist in manuscripts today. The list will not include the summaries of the first fourteen treatises of the so-called 'Canon', for they have been the object of recent scholarly attention, although they were not mentioned by Ḥunayn.[2] Following the list, one example will be explored in more detail – the *jawāmi'* of Galen's treatise on eye diseases.

1 Fuat Sezgin, *Medizin-Pharmazie-Zoologie-Tierheilkunde bis ca 430 H* (Leiden 1970) 140-50; Manfred Ullmann, *Die Medizin im Islam* (Leiden 1970) 65-67 and 343; Albert Dietrich, 'Medicinalia Arabica: Studien über arabische medizinische Handschriften in türkischen und syrischen Bibliotheken', *AAWGöttingen* 1966, no. 11; Gregor Schoeler, *Arabische Handschriften, Teil II* (Stuttgart 1990) 216-21; and I. Garofalo, 'La traduzione araba dei compendi alessandrini delle opere del canone di galeno. Il compendio dell' *Ad Glauconem*', *Medicina nei Secoli* 6 (1994) 329-48.

2 Elinor Lieber, 'Galen in Hebrew: the transmission of Galen's works in the mediaeval Islamic world', in *Galen: Problems and Prospects*, ed. Vivian Nutton (London 1981) 167-86; A. Z. Iskandar, 'An attempted reconstruction of the late Alexandrian medical curriculum', *Medical History 20* (1976) 235-38; and O. Temkin, 'Studies on Late Alexandrian Medicine, I: Alexandrian Commentaries on Galen's *De Sectis ad Introducendos*', *BHM 3* (1935) 405-30. The summaries of those treatises associated with the 'Canon' are currently being edited by Ivan Garofalo.

In addition to their association with the Alexandrians (stated in most, but not all, of the preserved copies), two distinctive visual didactic devices are associated with the *jawāmi'*. One of these is the presentation of material in synoptic tables.

The example illustrated in Fig. 1 is from a *jawāmi'* of Galen's treatise on materia medica (*De simplicium medicamentorum*) preserved in a twelfth or early thirteenth-century manuscript now in the Bodleian Library.[3] Immediately before the start of these synoptic tables there is a note (fol. 19b, bottom 4 lines) that apparently occurred in the exemplar used by the present unnamed copyist. In it the compiler of the epitome stated that 'I found seven chapters of the book, and so I decided to arrange them in branch-diagrams (*tashjīr*) and tables (*jadāwil*) so that the book would be a better abridgment, more worthy of being taken seriously, more attractive to see, easier to read, and not so boring'.

The term *tashjīr* refers to the second didactic device commonly found in the *jawāmi'* genre. An example can be seen in Fig. 2, from the same copy of the *jawāmi'* of Galen's treatise *De simplicium medicamentorum* as was illustrated in Fig. 1.

This branch-diagram format was usually called in Arabic *tashjīr*, meaning 'ramification', from a root meaning 'to plant with trees'. The method organized information into a diagrammatic form clearly delineating categories, divisions, and sub-divisions so that they could be easily remembered. Such a diagrammatic format as is illustrated in Fig. 2 was employed not just in Galenic summaries but also in some other early Arabic writings. For example, Ibn Māsawayh (d. 857), the teacher of Ḥunayn ibn Isḥāq, is recorded as using them, and the tenth-century writer Ibn Farī'ūn used them throughout his treatise on the classification of the sciences, which he titled *Jawāmi' al-'ulūm* ('The Compendium of Sciences').[4]

The origin of the technique of branch-diagramming is a matter of speculation – that is, whether it was a didactic tool originating in Alexandria in Late Antiquity or whether it arose within an Islamic context. There is persuasive evidence, however, that it was in use in the ninth century in the Arabic world. The technique was later introduced into the Latin West, where from the twelfth century it was frequently employed in discourses on the classification of the sciences or other scientific and philosophical topics.[5]

3 Oxford, Bodleian Library, Oriental Collections, MS Hunt. 600, item 2, folios 15a-35a. On the title page it is stated that the epitome was made by Ḥunayn ibn Isḥāq, although at the start of the treatise (illustrated in Fig. 2) it is said to be a *jawāmi'* of Galen's book on simple medicaments translated by Ḥunayn ibn Isḥāq. As will be shown below, Ḥunayn did record that he prepared a Syriac summary that was in his day translated into Arabic for one 'Ali ibn Yaḥyá. For a full description of the manuscript, see E. Savage-Smith, *A New Catalogue of Arabic Manuscripts in the Bodleian Library, Oxford, Volume 1: Arabic Manuscripts on Medicine and Related Topics* (Oxford, in press).

4 Ibn Farī'ūn, *Compendium of Sciences*: *Jawāmi' al-'ulūm* (Frankfurt am Main 1985). This is a facsimile of Istanbul, Topakapi Saray, Ahmed III MS 2768.

5 Modern scholars have called the Latin branch-diagrams 'dichotomies' to distinguish them from 'arbores' or tree-diagrams in which material was written in small cells arranged within the outline of a large tree having a trunk and root at the bottom. The earliest instance of an *arbor*-diagram occurs in a ninth-century copy of the *Etymologies* written in the seventh century by Isidore of Seville. While it is evident that in the ninth century branch-diagrams were being used in Arabic treatises (possibly continuing a now lost Alexandrian tradition) and tree-diagrams were being used in the Latin West, the two techniques of diagramming are sufficiently different as to suggest independent traditions. The Arabic method transferred to Europe, while the Latin one remained restricted to Western compositions. For the Latin tradition, see John E. Murdoch, *Album of Science: Antiquity and the Middle Ages* (New York 1984), 38-51.

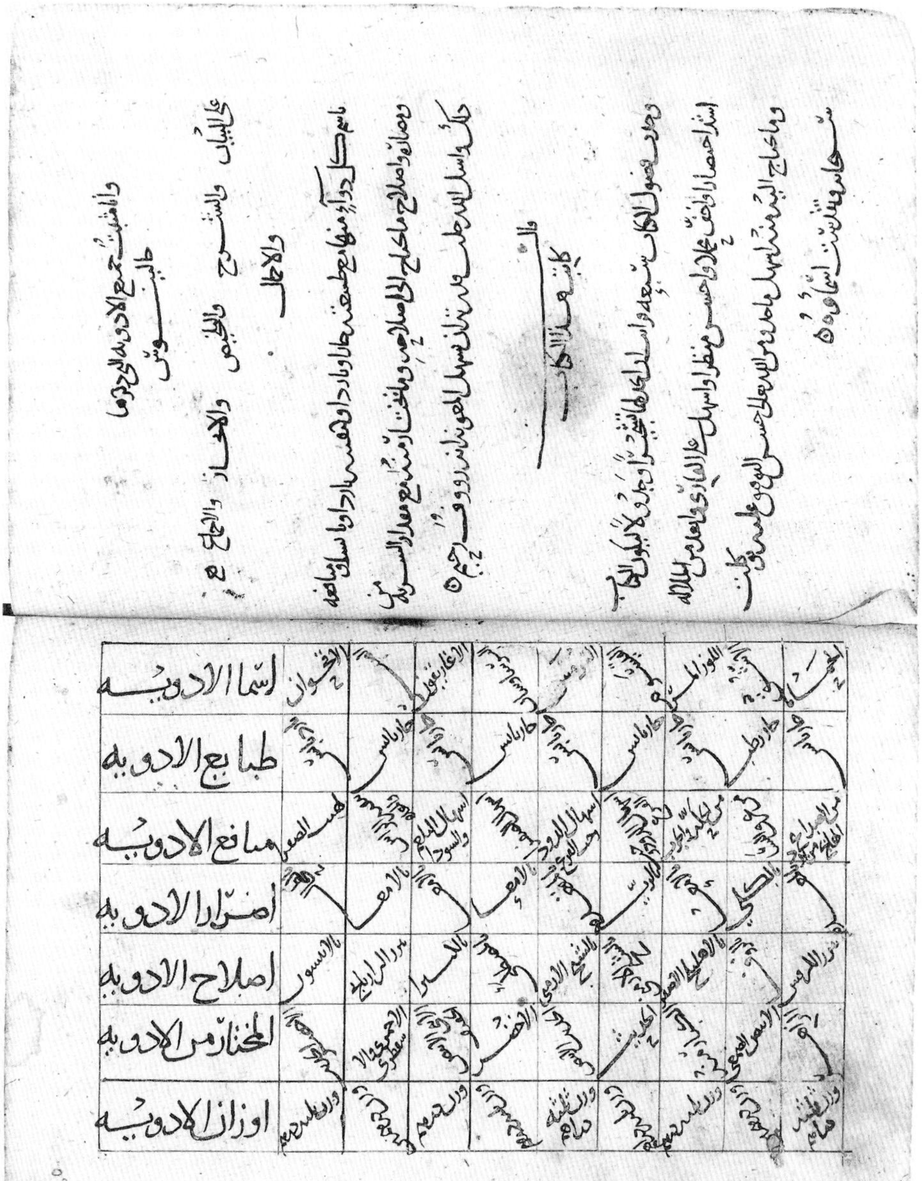

Figure 1: Synoptic tables in *Jawāmiʿ Kitāb Jālīnūs fī al-adwiyah al-mufradah* Oxford, Bodleian Library, Oriental Collections, MS Hunt. 600, fols 19b-20a (undated; 12-13th cent?). Reproduced by permission of The Bodleian Library.

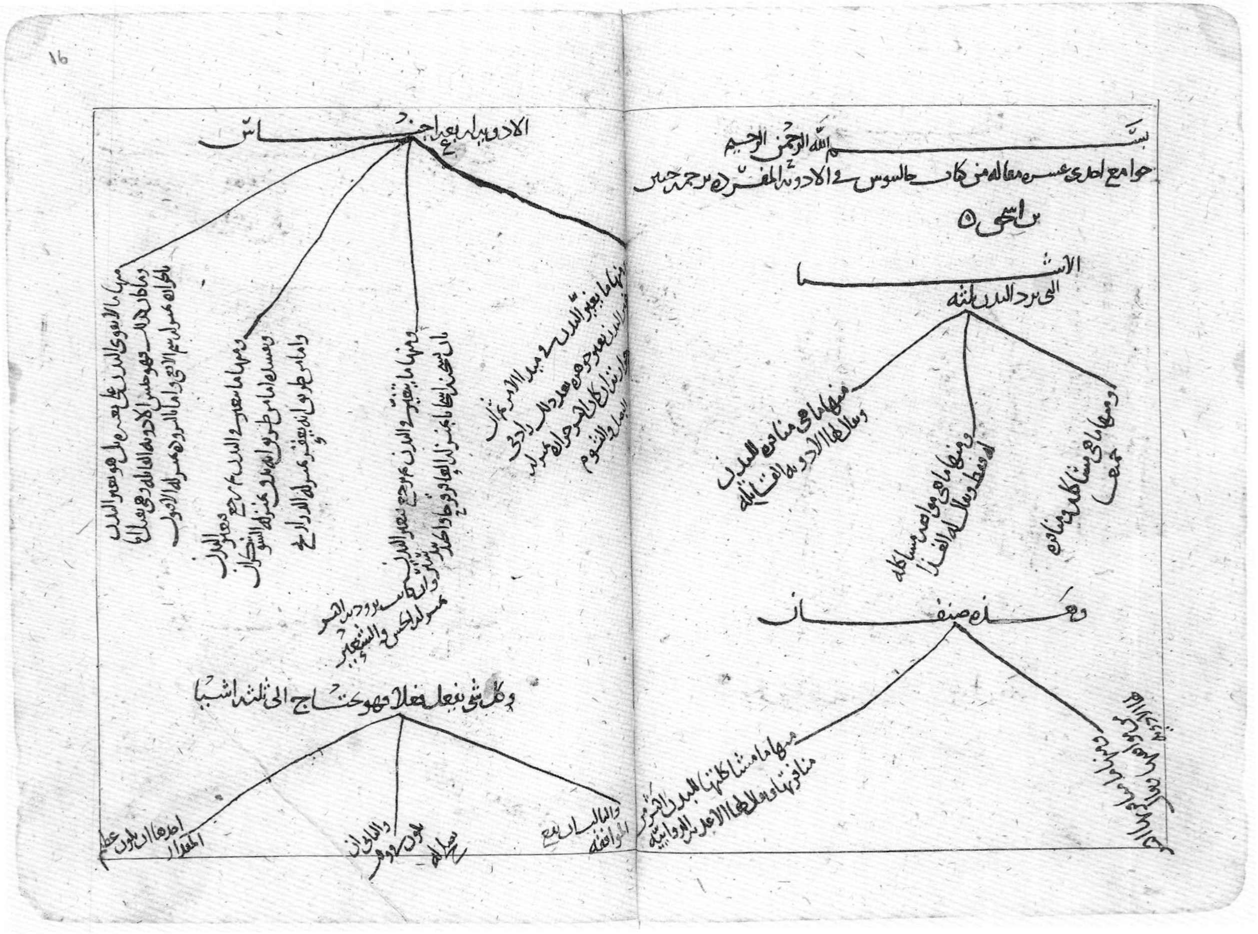

Figure 2: Branch-diagrams in *Jawāmi‘ Kitāb Jālīnūs fī al-adwiyah al-mufradah*, Oxford, Bodleian Library, Oriental Collections, MS Hunt. 600, fols 15b-16a (undated; 12-13th cent?).

Similar diagramming techniques were recently noted as occurring in a manuscript now in Vienna (*Codex medicus graecus* 16) which contains Greek summaries of five Galenic treatises. This manuscript was examined in detail by Beate Gundert as a witness to the Alexandrian medical tradition of the sixth century AD.[6] The manuscript, acquired in Constantinople by Augerius von Busbeck (Ogier Ghiselin de Busbecq, d. 1592), is undated but estimated to be from the thirteenth to fifteenth centuries. The discovery of Greek summaries of five Galenic treatises associated with the so-called 'Canon' is very important, and is made even more so by the occurrence of branch-diagrams of the same type as found in some of the Arabic *jawāmi'*. In this particular Greek manuscript, however, the Greek texts appear to differ widely from the Arabic versions of the *Summaria* for the same five treatises – though here perhaps we must await the study and edition by Garofalo of the Arabic *jawāmi'* of Galen's 'Therapeutics for Glaucon'.[7] In addition, it would appear that branch-diagrams were not used in the Arabic manuscripts of the summaries of the 'Canon' texts, though this requires further investigation. Rather, it is the *jawāmi'* of other Galenic treatises that most often employ the technique.

Since, by the twelfth century, branch-diagrams were not infrequently used in Europe to summarize and present a text, it is not impossible that these Greek summaries, presented in a diagrammatic format and preserved in a manuscript estimated to be from the thirteenth to fifteenth centuries, are a late medieval Byzantine product rather than representing a sixth-century Alexandrian prototype. I am, however, inclined to think that this didactic technique may have been an Alexandrian invention – though unfortunately sources such as Paul of Aegina, working in Alexandria in the seventh century, or the Galenic commentators Stephanus and John the Grammarian or John of Alexandria, are regrettably silent on the subject. If indeed we accept the proposition that this *tashjīr* format reflects an Alexandrian origin, then we must give more attention to the considerable number of synopses of Galenic treatises that are preserved in Arabic in this format – and most (if not all) of these are summaries of treatises other than those considered part of the so-called 'Canon' of sixteen books.

When considering these numerous Galenic summaries as a whole, we must first observe that Ḥunayn (who died either in AD 873 or 877) was surely in a position to know of them, had they circulated in his day. Yet he mentions only one: a summary of 'On the therapeutic method' (*De methodo medendi*) which he says was translated into Syriac, presumably from the Greek, by one Ibrāhīm ibn al-Ṣalt.[8] But Ḥunayn does not associate this summary with the Alexandrians, but rather attributes this to Galen himself, and he mentions no other summaries of the treatises associated with the core group of books or 'Canon' taught by the Alexandrians.[9] He does go

6 Beate Gundert, 'Die *Tabulae vindobonenses* als Zeugnis alexandrinischer Lehrtätigkeit um 600 n. Chr.', in *Text and Tradition: Studies in Ancient Medicine and its Transmission, presented to Jutta Kollesch*, ed. Klaus-Dietrich Fischer, Diethard Nickel, and Paul Potter, (Studies in Ancient Medicine, 18) (Leiden 1998) 91-144. The manuscript contains Greek summaries of *De sectis, Ars medica, De pulsibus ad tirones, Ad Glauconem*, and the group of four short tracts on the causes and symptoms of diseases.

7 See Garofalo, 'Traduzione' (above, n. 1).

8 G. Bergsträsser, *Ḥunain ibn Isḥāq über die syrischen und arabischen Galen-Übersetzungen*, Abhandlungen für die Kunde des Morgenlandes, 17, 2 (Leipzig 1925) *Risālah* 70.

9 I. Garofalo, 'Una nuova opera di Galeno: la *synopsis* del *De methodo medendi* in versione araba', forthcoming. The manuscript, Princeton Garrett 1075 contains the Arabic Alexandrian summaries of the first seven books, and Galen's own synopsis of the last seven and a portion of Book 7. Given that Ḥunayn's list deals with Galen's writings, perhaps it is not surprising that he does not refer to others' summaries of them.

on, however, to record that he *himself* made summaries (*jawāmi'*) of eleven other Galenic treatises, which he says he prepared in Syriac. Of these eleven Syriac *jawāmi'*, some, he states, were translated into Arabic by 'Īsā ibn Yaḥyā, while others were translated either by Ḥunayn himself or his nephew Ḥubaysh for one Muḥammad ibn Mūsá or one 'Alī ibn Yaḥyá.[10] In two instances,[11] Ḥunayn stated that he prepared the summaries (*jawāmi'*) in a tabular form (*ṭarīq al-taqāsīm* or *'alá al-taqsīm*), possibly meaning synoptic tables such as we saw earlier in Fig. 1. Unfortunately, he does not use the term *tashjīr*, so we are not certain that he either knew of or used himself the branch-diagram format. He also stated that in two instances he summarized the treatises in a question-and-answer format, though no *jawāmi'* preserved today are in a question-and-answer form.

In the extant copies of the Galenic *jawāmi'*, Ḥunayn's name is often given as the Arabic translator. A close association with Ḥunayn's circle is also implicit in a marginal annotation occurring in an undated copy (possibly eleventh to thirteenth century) of a *jawāmi'* of *De locis affectis*.[12] In it the copyist states that his exemplar was corrected in the handwriting of al-Azraq, the copyist (*warrāq*) for Ḥunayn.

In order to develop a fuller picture of the origin and role of Arabic summaries of Galenic treatises, and their relationship to Greek and Syriac material, we need to review the relationship between the so-called Alexandrian 'Canon' and the various Galenic *jawāmi'* that are known to be preserved today. According to Ḥunayn, there were twenty-four Galenic treatises that were read 'by the Alexandrians'. But sixteen of these formed four groups of four related treatises each; therefore it is possible to count the total number of treatises as twelve rather than twenty-four. Ḥunayn does not refer to these as the Alexandrian 'Canon' nor does he ever mention a group of 'sixteen books'. Moreover, according to Ḥunayn, all the treatises were read in their entirety, except for the group of four on pulse, where only the first book of each treatise was read, and 'On the therapeutic method', where only the last eight books were read. Ḥunayn makes no mention of synopses used by the Alexandrians.

Later sources, however, refer to 'sixteen books' read by the Alexandrians, in which the first four were counted separately (rather than as one group as Ḥunayn had described) and an additional treatise was added. This additional treatise could be either *De sanitate tuenda* or *De demonstratione*, depending upon one's source of information. At this point, however, we are still talking about the reading of complete, unabridged, treatises.[13] All this has been well rehearsed by other scholars, and very succinctly so by Elinor Lieber twenty years ago.[14] Modern scholars have sometimes tended to confound this group of sixteen books, which

10 See the list given below, and also Bergsträsser (n. 8 above.), and Rainer Degen, 'Galen im Syrischen: eine Übersicht über die syrische Uberlieferung der Werke Galens', in *Galen: Problems and Prospects* (above, n. 2) 131-66.

11 Summaries of the treatise on marasmus and the treatise on unnatural growths.

12 Berlin, Staatsbibliothek, MS Or. oct. 1122, fol. 194b; see Schoeler, *Arabische Handschriften* (above, n. 1), 216-21, no. 203, with illustration in Plate 8. This manuscript contains *jawāmi'* of the four short tracts on anatomy (treated as one treatise), the four tracts on the symptoms and causes of diseases (considered as one treatise), *De nat. fac.*, and *De locis affectis*.

13 Moreover, in these later lists of sixteen books forming the Alexandrian 'Canon', all the books of the four treatises on pulse (that were counted as a single group) and all the books of *De methodo medendi* were said to be read by the Alexandrians, and not selected books as Ḥunayn had recorded.

14 See above, n. 2.

came to be called the Alexandrian 'Canon', with the 'Alexandrian summaries' (*jawāmi' al-Iskandarānīyīn*), which are yet another, but different, link in the chain.

The earliest mention we have of summaries of these treatises is a ninth-century scholar from Edessa, Isḥāq ibn 'Alī al-Ruhāwī, who wrote an important treatise on medical ethics. He stated that the Alexandrians 'organized the books of Galen as sixteen books, abridged them as compendia in an effort to shorten and summarize them, and taught them in the *skolê*, the place where they used to teach'.[15] The origin of the summaries has been much debated but still remains unsettled. Temkin, Peterson, Westerink, Manetti, Palmieri, Garofalo, and Gundert have all made significant contributions to this discussion, and Garofalo is currently editing some of the summaries of the treatises forming the so-called 'Canon'.[16]

The name Yaḥyá al-Naḥwī is one which repeatedly occurs in association with the preserved *jawāmi'*. He is named as the author of summaries of the 'Canon' of sixteen books which are preserved in a copy completed in 1218,[17] and as the author of a *jawāmi'* of Galen's *De usu partium* of which only fragments are preserved.[18] The name is one that also figures in medieval bio-bibliographical writings in which compilers of the *Summaria Alexandrinorum* are suggested, but it is difficult to determine the identity of the person intended. At least three or four people were known in Arabic as Yaḥyá al-Naḥwī: (1) the sixth-century Aristotelian commentator John Philoponus; (2) the author of Greek commentaries on Galen including a commentary on *De usu partium*, translated into Arabic by Ibn Zur'ah in the tenth century, who probably was a medical writer living in Alexandria in the late sixth or early seventh century and possibly the same person who wrote a medical history; (3) the author of a summary of what is stated to be the sixteen books of the Alexandrians (but in fact contains only fourteen synopses), who might also be the same as (4) the author of a summary (*jawāmi'*) of *De usu partium*, the latter probably being an Arabic-speaking Yaḥyá al-Naḥwī.[19]

15 Trans. by Dimitri Gutas, *Greek Thought, Arabic Culture: The Graeco-Arabic Translation Movement in Baghdad and Early 'Abbasid Society (2nd-4th/8th-10th centuries)* (London 1998) 93. For al-Ruhāwī and his list, see above, n. 2.

16 Temkin, 'Studies', above, n. 2; D.W. Peterson, *Galen's 'Therapeutics to Glaucon' and its Early Commentaries*, diss., Baltimore, Johns Hopkins University,1974; D. Manetti, 'P.Berol. 11739A e i commenti tardoantichi a Galeno', in *Tradizione e ecdotica dei testi medici tardoantichi e bizantini*, ed. A. Garzya (Naples, 1992), 211-35; N. Palmieri, 'Survivance d'une lecture alexandrine de l'*Ars medica* en latin et en arabe', *Archives d'histoire doctrinale et littéraire du moyen âge*, LX (1993), 57-102; L. G. Westerink, *Agnellus of Ravenna. Lectures on Galen's 'De sectis'*, Arethusa Monographs 8, (Buffalo 1981): Garofolo, 'Traduzione', above, see n. 1; Gundert , 'Tabulae', above, n. 6.

17 London, BL, MS Arundel Or. 17. The colophon is unclear, but seems to read 615/1218 or 625/1227. The manuscript is stated to be a *talkhīṣ* of the sixteen books, but only fourteen are actually in the copy. The two missing are: '*De methodo medendi* and 'The large pulse'.

18 Paris, BnF, fonds arabe 2853 (fragment given at the end of the fourteenth book of the Arabic translation of *De usu partium*) and Bethesda, MD, National Library of Medicine, MS A 30.1, folio 209a. For a description of the latter manuscripts, see the section 'Translations of Earlier Sources: Galen' in the on-line catalogue, E. Savage-Smith, *Islamic Medical Manuscripts at the National Library of Medicine* (http://www.nlm.nih.gov/hmd/arabic).

19 Ibn al-Qifti, *Ta'rīkh al-ḥukamā'*, ed. J. Lippert (Leipzig 1903), 354-57; Ibn Abī Uṣaybi'ah, *'Uyūn al-anbā' fī ṭabaqāt al-aṭibbā'*, 2 vols, ed. A. Müller (Cairo and Königsberg 1882-84), II, 103-09; Ullmann, *Medizin* (above, n. 1), 89-91; Max Meyerhof, 'Joannes Grammatikos (Philoponos) von Alexandrien und die arabische Medizin', *MDAI (Cairo) 2* (1932) 1-21; E. Savage-Smith, *Galen on Nerves, Veins and Arteries: A critical edition and translation from the Arabic* (diss. Univ. of Wisconsin-Madison, 1969), 17-29; and V. Nutton, 'John of Alexandria Again: Greek Medical Philosophy in Latin Translation', *Classical Quarterly* 85 (1991) 509-19.

According to bio-bibliographical sources there were other summaries circulating in the ninth century, most importantly those attributed to Thābit ibn Qurrah (d. 901),[20] Ibn Abī al-Ash'ath (d. 970),[21] and Ibn Zur'ah (d. 1008),[22] and copies of some of these are known to be extant today. A comparison of the extant *jawāmi'* attributed to these figures with anonymous versions would be an essential element in eventually understanding the origin and development of Arabic Galenic summaries.[23]

The following list begins with the twelve summaries mentioned by Ḥunayn. The synopsis of *De methodo medendi* is the only one of the so-called 'Canon' of sixteen books included in Ḥunayn's account of summaries. Following the common Latin title for the Galenic treatise, the description of the summary provided by Ḥunayn is then given, accompanied by a reference to the pertinent section of his *Risālah* in which he discussed the Galenic treatises.[24] Versions of the *jawāmi'* that are known to be preserved are then listed, as well as additional *jawāmi'* of the treatise either given in bio-bibliographical sources or extant in manuscripts. An asterisk (*) before a manuscript indicates that the item has not been inspected and cannot therefore be classified with certainty. If no manuscript is given, then none are as yet known to be preserved. Those Galenic treatises specifically mentioned by Ḥunayn as having had a synopsis made of them are then followed by those treatises for which *jawāmi'* are known to be preserved (other than the remaining 'Canon' of treatises) even though such a summary is not mentioned by Ḥunayn.

De methodo medendi

1. Galen's own synopsis translated into Syriac by Ibrāhīm ibn al-Ṣalt (*Risālah* 70). Arabic translation, books 7 (incomplete) and 8-14. Princeton, University Library, Garrett Coll., MS 1075 (no. 1 G), item 4.
2. *Jawāmi' Kitāb ḥīlat al-bur'*
   *Istanbul, Topkapi Saray, Ahmed III, MS 1043
   *Florence, Biblioteca Laurentiana, cod. arab. 235, item 15
3. *Jawāmi' Kitāb ḥīlat al-bur'* [different content from that of No. 2]
   Oxford, Bodleian Library, MS Hunt. 600, fols. 1b-45 (*tashjīr* format and tables)
   Dublin, Chester Beatty Library, Arabic MS 4001, fols. 9b-14b
   *Tehran, Dānishgāh MS 5217, fols. 95b-102b
   *Tehran, Majlis, MS 6037, item 2
4. *Jawāmi'/ Ikhtiṣār* by Thābit ibn Qurrah

20 He is recorded as having made *jawāmi'* of *De pulsibus ad tirones* (the four introductory tracts on pulse), *De elementis*, and *De diebus decretoriis*. His *jawāmi'* of the four tracts on the symptoms and causes of disease (*Jawāmi' Kitāb Jālīnūs fī aṣnāf al-amrāḍ*) is preserved in Istanbul, Süleymaniye, Ayasofya MS 3631, fols. 62a-65a. See the list below and Sezgin, *Medizin-Pharmazie* (above, n. 1) 261-62.

21 For Ibn al-Ash'ath, see Ullmann, *Medizin* (above, n. 1) 138-39. A copy of his summary of *De temperamentis* is preserved in Tokyo, Daiber MS 133, folios 154b-167a.

22 His *jawāmi'* of *De elementis, De temperamentis, De facultatibus naturalibus,* and Galen's four introductory tracts on anatomy are extant in Tehran, Majlis MS 6037. See Sezgin, *Medizin-Pharmazie*, (above, n. 1) 147-48.

23 There were additional summaries produced in the tenth and eleventh centuries: al-Nīlī (d. 1029) whose summary (*talkhīṣ*) of Galen's commentary on the *Aphorisms* of Hippocrates is preserved in Oxford, Bodleian Library, Oriental Collections, MS Hunt. 359, fols. 1a-30a; Abū al-Faraj ibn al-Ṭayyib (d. 1043) made a series of summaries, which he called *thimār*, of eight treatises comprising the 'Canon'; and Ibn Rushd (d. 1198) composed a summary of the book on temperaments and another of *De febr. diff.*

24 The numbers refer to sections in the Arabic edition and German translation given by Bergsträsser, *Ḥunain ibn Isḥāq* (above, n. 8).

De simplicium medicamentorum facultatibus

1. A summary prepared by Ḥunayn in Syriac and translated into Arabic for ʻAlī ibn Yaḥyá (*Risālah* 53).
2. *Jawāmi' Ma'ānī al-khams al-maqālāt al-ūlā min Kitāb Jālīnūs fī quwā al-adwiyah al-mufradah al-mansūqah 'alā ṭarīq al-mas'alah wa-al-jawāb* [Ḥunayn's translation of his Syriac summary]
   *Istanbul, Nuruosmaniye, MS 3555
3. *Jawāmi' Kitāb Jālīnūs fī al-adwiyah al-mufradah*
   Oxford, Bodleian Library, MS Hunt. 600, fols. 15a-35b (*tashjīr* format)
   *Florence, Biblioteca Laurentiana, cod. arab. 235, item 17

De marcore / De marasmo

1. A Syriac summary in tabular format (? *ṭarīq al-taqāsīm*) by Ḥunayn which was translated into Arabic by ʻĪsá (*Risālah* 72).
2. *Jawāmi' Kitāb Jālīnūs fī al-dhubūl* by Thābit ibn Qurrah
   Oxford, Bodleian Library, MS Marsh 215, fols. 78b and 80a-85b
   Istanbul, Süleymaniye, Ayasofya, MS 3631, fols. 38b-45a

De tumoribus praeter naturum

A Syriac summary in tabular format (? *'alá taqsīm*) by Ḥunayn (*Risālah 57).*

De venarum arteriarumque dissectione

A Syriac summary by Ḥunayn that was translated into Arabic for Muḥammad ibn Mūsá (*Risālah* 10).

De atra bile

1. A Syriac summary by Ḥunayn translated into Arabic by ʻĪsá (*Risālah* 64).
2. *Jawāmi' Kitāb fī mirrat al-sawdā'*
   Istanbul, Süleymaniye, Ayasofya, MS 3716, fols. 284b-287b

In Hippocratis epidemiarum librum VI commentarii. VIII

A Syriac summary in question-and-answer format by Ḥunayn, translated into Arabic by ʻĪsá ibn Yaḥyá (*Risālah* 95).

In Hippocratis librum de acuturum victu commentarii. IV

A Syriac summary in question-and-answer format by Ḥunayn, partially translated into Arabic by ʻĪsá ibn Yaḥyá (*Risālah* 92).

In Hippocratis de capitis vulneribus librum commentarius

A Syriac summary by Ḥunayn (*Risālah* 94).

De Hippocratis scriptis genuinis

A Syriac summary by Ḥunayn, translated into Arabic for ʻAlī ibn Yaḥyá (*Risālah* 104)

In Hippocratis de officina medici commentarii III

A Syriac summary by Ḥunayn, translated into Arabic by Ḥubaysh for Muḥammad ibn Mūsá (*Risālah* 98).

De victu attenuante

1. A Syriac summary by Ḥunayn, translated into Arabic by ʿĪsá ibn Yaḥyá (*Risālah* 75).
2. *Jawāmiʿ Kitāb Jālīnūs fī al-tadbīr al-mulaṭṭif*
   Oxford, Bodleian Library, MS Marsh 663, pp. 265-271, said to be 'made by Ḥasan from Ḥunayn's translation'
   *Istanbul, Süleymaniye, Ayasofya, MS 3631, fols. 110b-115a, also stated to be in Ḥunayn's translation
3. *Ikhtiṣār*, anonymous (cited by Ibn Abī Uṣaybiʿah)

De victus ratione in morbis acutis ex Hippocratis sententia
Jawāmiʿ Kitāb tadbīr al-amrāḍ al-ḥāddah ʿalá raʾy Buqrāṭ, by (?) Thābit ibn Qurrah
*Istanbul, Süleymaniye, MS 3631, fols. 45b-55a

De inaequali intemperie
Jawāmiʿ Kitāb Jālīnūs fī sūʾ al-mizāj al-mukhtalif, by Thābit ibn Qurrah
*Istanbul, Süleymaniye, MS 3631, fols. 34a-38b

De purgantium medicamentorum facultate
Jawāmiʿ Kitāb Jālīnūs fī quwa al-adwiyah al-mushilah, by Thābit ibn Qurrah
*Istanbul, Süleymaniye, MS 3631, fols. 27a-33b

De septimestri partu
Jawāmiʿ Kitāb Jālīnūs fī al-mawlūdīn li-sabʿat ashhur, by Thābit ibn Qurrah
*Istanbul, Süleymaniye, MS 3631, fols. 58b-63a

De usu partium
Jawāmiʿ Kitāb Jālīnūs fī manāfiʿ al-aʿḍāʾ, by Yaḥyá al-Naḥwī
Paris, BnF, fonds arabe 2853 (fragments)
Bethesda, MD, National Library of Medicine, MS A 30.1 (fragments)

De uteri dissectione
Jawāmiʿ Kitāb fī tashrīḥ al-raḥim, by Thābit ibn Qurrah
*Istanbul, Süleymaniye, MS 3631, fols. 55a-58b

De urinis

1. *Jawāmiʿ Kitāb fī al-bawl wa-dalāʾil*
   Oxford, Bodleian Library, MS Turk. e. 33, fols. 195b-204a (*tashjīr* format)
   London, British Library, OIOC, MS Or. 5862 item 7 (*tashjīr* format)
   Vatican, Biblioteca Apostolica, cod. arab. 1662, fols. 69a-83b
   Bethesda, MD, National Library of Medicine, MS A84 item 3
   Dublin, Chester Beatty Library, MS Arabic 4884 item2
   *Cairo, Dār al-Kutub, MS 1783 *ṭibb*
   *Mosul, MS 153 item 7
2. A Judeo-Arabic synopsis
   *Vatican, Biblioteca Apostolica, cod. heb. 369
3. A Hebrew summary
   *Vienna, Nationalbibliothek, hebr. cod. 174

De urinis [a different version]

Jawāmi' Jālīnūs fī dalā'il al-bawl wa-ṣifat al-amrāḍ, in 17 chapters

Oxford, Bodleian Library, MS Marsh 663, pp. 271-4

Oxford, Bodleian Library, MS Turk. e. 33, fols. 178b-179b

De cura morborum

Jawāmi' Jālīnūs fī shifā' al-amrāḍ

London, British Library, OIOC, MS Add. 23407, fols. 129a-156b

De catharticis medicamentis

Jawāmi' Jālīnūs fī al-adwiyah al-munqiyah

Oxford, Bodleian Library, MS Marsh 215, fols. 80a-85b

*Istanbul, Süleymaniye, Ayasofya, MS 3631, fols. 27b-33b

De morbis oculorum

Jawāmi' Kitāb Jālīnūs fī al-amrāḍ al-ḥādith fī al-'ayn

St Petersburg, Institut narodov Azii, MS C 875, fols. 70a-76b

Cairo, MS Ṭibb Taymur 100, pp. 2-9

Dublin, Chester Beatty Library, MS Arabic, fols. 141a-156b

Tehran, Dānishgāh MS 4914, fols. 440a-441b

Let us now examine in greater detail one of the *jawāmi'* that are not based on a text associated with the 'Canon' and one whose original Greek text (if indeed it is based on a genuine Galenic treatise) is now lost – the *Jawāmi' Kitāb Jālīnūs fī al-amrāḍ al-ḥādith fī al-'ayn* or 'The Summary of Galen's Book on diseases that occur in the eye'. It is preserved in four manuscript copies.[25]

The earliest copy (St Petersburg, Institut narodov Azii, MS C 875) was transcribed between 1151 and 1160 AD by 'Abd al-Raḥmān ibn Ibrāhīm ibn Sālim ibn 'Ammār al-Maqdisī al-Ansāṛī *al-mutaṭabbib* [the physician].[26] In this copy there are no branch-diagrams, but there are some circular diagrams in which the names of ocular disorders are written along the radii of a circle. The second oldest manuscript (Cairo, Dār al-Kutub, MS Ṭibb Taymūr 100) was copied between 1195 and 1197, directly from the St Petersburg manuscript by 'Abd al-Raḥīm ibn Yūnus ibn Abī al-Ḥasan al-Anṣārī, who states that he made the copy for himself.[27]

25 Cairo, Dār al-Kutub, MS Ṭalat 616 *ṭibb* is a modern photocopy of the MS Ṭibb Taymūr 100, although it is cited as a separate treatise by Sezgin, *Medizin-Pharmazie* (above, n. 1) 102, and by Ṣalaḥ al-Dīn al-Munajjid, 'Maṣādir jadīdah 'an ta'rīkh al-ṭibb 'inda al-'arab,' *Revue de l'Institut des Manuscrits Arabes* 5 (1959) 288.

26 St Petersburg, Institut narodov Azii, MS. C 875, folios 70a-76b. There are ll items in the manuscript, copied between 1151 and 1160 (545-554 H). For a description of the manuscripts, see Max Meyerhof, 'New Light on the Early Period of Arabic Medical and Ophthalmological Science', *Bulletin de la Sociéte ophtalmologique d' Égypte* (1961) 25-37. The copy was begun in approximately the same year as the earliest dated copy of a *jawāmi'* based on a treatise in the so-called 'Canon' – that is, London, BL, MS Or. 9202, which contains four summaries of parts of the 'Canon' and was copied not long before 1152,when the third of three owners received it from his father; MS Or. 9202 also has an important collation note stating that the volume was read before and corrected by Ibn al-Tilmidh (d. 1165), a prominent physician of Baghdad.

27 Cairo, Dār al-Kutub, MS Ṭibb Taymūr 100, item 1, pp. 2-9, copied between 592 and 594 H [1195-7 AD] . The copyist states that he made the copy from one made by al-Ḥakīm' Abd al-Raḥmān ibn Ibrāhīm ibn Sālīm ibn 'Ammār al-Anṣārī al-Maqdisī, sometimes called *al-kaḥḥāl* (the oculist), in the year 545 H [1151]. The manuscript consists of 452 numbered pages, and is a collection of ophthalmological treatises all transcribed by the same hand.

The third copy, now in Dublin (Chester Beatty Library, Arabic MS 3425), was completed in the month of Rajab 834 (Mar-Apr 1431) by Aḥmad ibn Muḥammad ibn Ḥusayn al-Ṭūsī *mahtidan* [by lineage] al-Ḥarawī *muwalladan* [by birth].[28] It employs a large number of branch-diagrams of varying forms. Figure 3 illustrates the opening leaves of the treatise. The copy is of particular interest in that Persian glosses were incorporated into the text at various points by the copyist. The fact that the Dublin copy has Persian glosses that have been incorporated into the text suggests that at least this one *jawāmi'* had considerable currency in the Persian-speaking world before the early fifteenth century, when it was transcribed.

The fourth and most recent copy is now in Tehran (Dānishgāh MS 4914) and is an undated copy of probably the seventeenth century which does not use branch-diagrams.[29] All four copies also have diagrams of the eye and its relationship to the brain, accompanied by lists of the various tunics and parts of the eye. One example is shown in Figure 4.

The treatise consists basically of an enumeration of ninety-one eye diseases and symptoms, intermixed with many Greek terms (in transliteration) and ending with a listing of the parts of the eye with an accompanying diagram of the visual system. Though it is stated explicitly to be a *jawāmi'* of Galen's book on diseases that occur in the eye, does it in fact represent a summary of genuine Galenic work?[30] Since Galen's only known treatise on ophthalmology is now lost in the Greek, can we use it to expand our knowledge of Galen's ideas regarding eye diseases?

Galen in 'On his own books'[31] says that when he returned from Rome at the age of 37 he had composed three books, one of these a small book 'On the diagnosis of diseases of the eyes' (*Peri tôn en ophthalmois pathôn, mikron*), which was written for a young man who treated the eyes.[32] Certainly a simple mnemonic guide to the various ocular disorders, as represented by the extant Arabic epitome, could be of use to a novice oculist. This small tract on eye complaints is the only treatise devoted solely to ophthalmological matters that Galen lists amongst his compositions in the lists as we have them today.

Whether Galen's Greek treatise was actually ever translated into Arabic is uncertain, though there was a Syriac translation made in the sixth century, according to Ḥunayn ibn Isḥāq. Ḥunayn gave the title of Galen's treatise as *Kitāb fī dalā'il 'ilal al-'ayn* ('on symptoms of eye diseases'), and, following Galen, he says that Galen composed it during his youth for a young oculist (*kaḥḥāl*). Ḥunayn adds that it was translated into Syriac by Sergius of Rēsh'aynā (d. 536 AD), and that while a Greek manuscript of it was available to him (Ḥunayn), he did not have time to translate it.[33] No mention is made of a summary of the treatise being available.

28 Dublin, Chester Beatty Library, Arabic MS 3425, item 2, folios 141a-156b, copied by the same scribe as made the next item in the volume, which was completed in Rajab 834 (Mar-Apr 1431) by Aḥmad ibn Muḥammad ibn Ḥusayn al-Ṭūsī.

29 Tehran, Dānishgāh MS 4914, fols. 440a-441b, undated (c. 11/17th century).

30 See Ullmann, *Medizin* (above, n. 1), 56 no. 87; Sezgin, *Medizin-Pharmazie* (above, n. 1) 101-02; and M. Steinschneider, *Die arabischen Übersetzungen aus dem Griechischen* (repr., Graz 1960) p. 349 no. 77.

31 Galen, *Scripta minora*, ed. I. Müller (Leipzig, 1891) II, $97_{12\text{-}16}$. See V. Nutton, 'Galen and Medical Auto-biography', *Proceedings of the Cambridge Philological Society*, n.s. 18 (1972) 50-62.

32 *Ophthalmous therapeuonti neaniskôi*, where the use of the term *therapeuonti* might suggest that he was not a physician but an empirically trained oculist.

33 Bergsträsser, *Ḥunain ibn Isḥāq* (above, n. 8), *Risālah* 54.

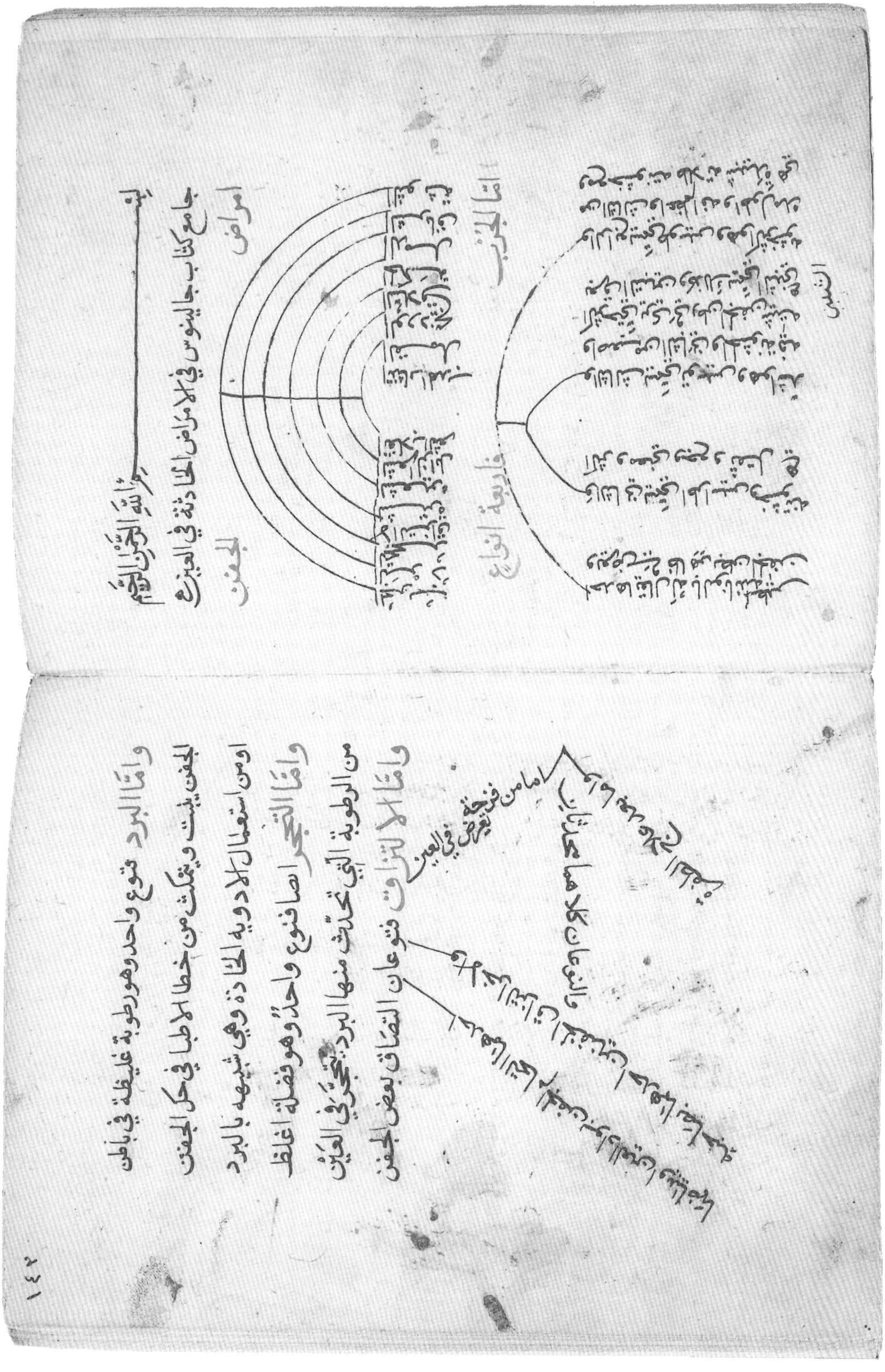

Figure 3: The opening of *Jawāmiʿ Kitāb Jālīnūs fī al-amrāḍ al-ḥādith fī al-ʿayn*, Dublin, Chester Beatty Library, Arabic MS 3425, fols. 141b-142a. Copy completed in Rajab 834 H (March-April 1431). Reproduced by kind permission of the Trustees of the Chester Beatty Library

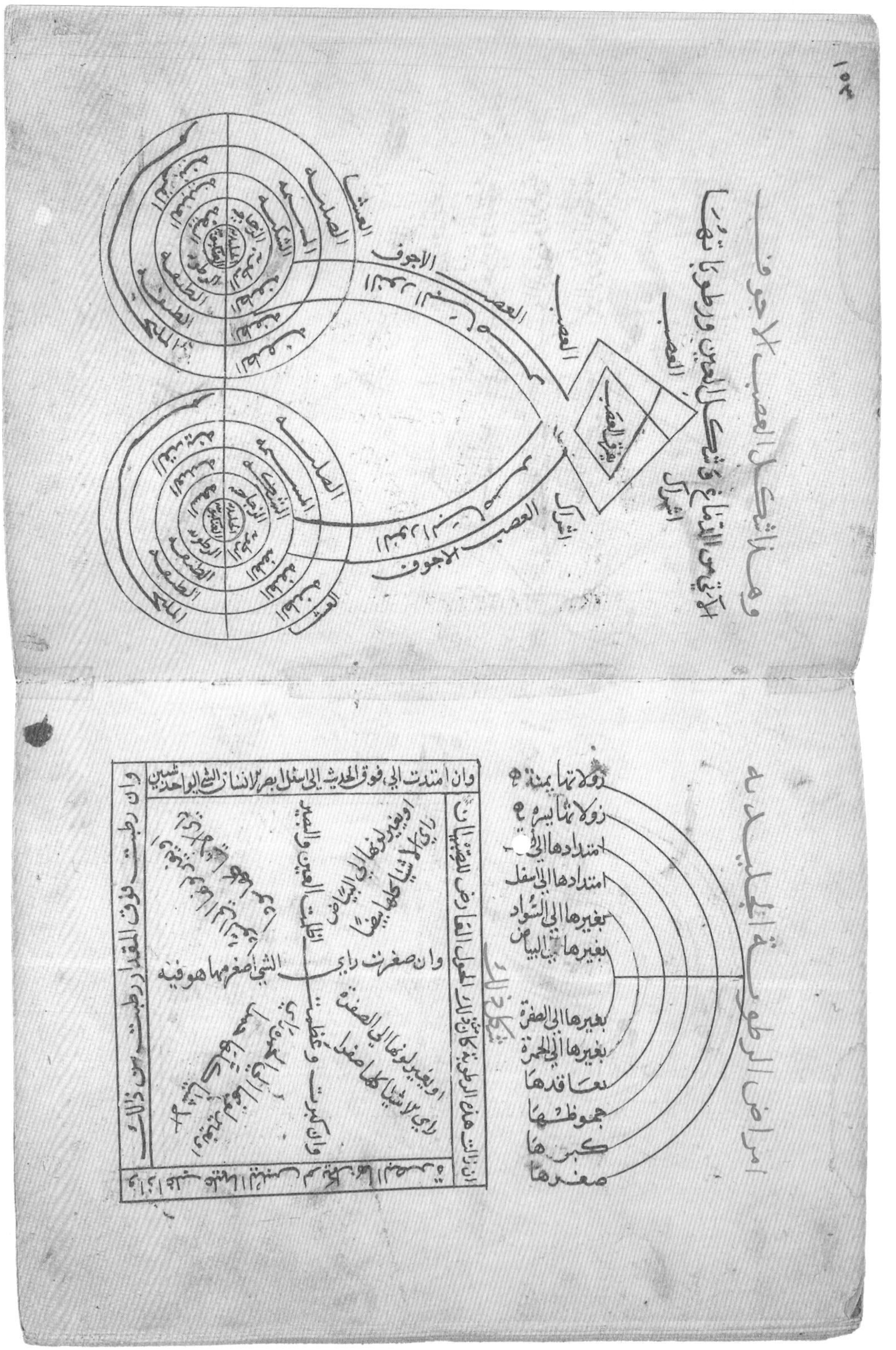

Figure 4: A diagram of the visual system from *Jawāmi' Kitāb Jālīnūs fī al-amrāḍ al-ḥādith fī al-'ayn*, Dublin, Chester Beatty Library, Arabic MS 3425, fols. 152b-153a. Copy completed in Rajab 834 H (March-April 1431).

Aḥmad ibn Abī Ya'qūb, known as al-Ya'qūbī, the noted historian and geographer who died in 897, gave a catalogue of Galen's writings.[34] There the title is given as *'Alāmāt al-'ayn* 'symptoms of the eye', using a different word for symptom than did Ḥunayn. It is not clear whether he actually saw these books, or is relying on an informant or taking them from citations in other unspecified writings.[35]

Ibn al-Haytham, who worked in Egypt and died in or after 1040, enumerated thirty books of Galen that he considered important (and presumably to which he had some access). These included Galen's 'Discourse on the diseases of the eye' (*Kalām fī amrāḍ al-'ayn*).[36] There is a current debate going on whether the biographies of two separate Ibn al-Haytham's (one a physician and one a mathematician) have been conflated into one.[37] But that is irrelevant for our purposes, the point being that the Galenic treatise was being cited, though by a slightly different title.

Ibn Abī Uṣaybi'ah, an historian and physician who died in Syria in 1270, essentially repeats the statement of Ḥunayn ibn Isḥāq, almost word for word, giving the same title, *Kitāb fī dalā'il 'ilal al-'ayn*.[38]

Thus there are four Arabic testimonies to an Arabic equivalent for the title of Galen's original tract on the diagnosis of eye diseases. Note that the title refers to the original treatise and not to a summary, which is not mentioned in the medieval bibliographic sources. The Arabic titles employ a variety of synonyms for the term 'symptom', but basically they all state that Galen wrote a treatise on symptoms of eye diseases, which would more or less correspond to Galen's own title of a treatise on the diagnosis of eye diseases. How available it might have been to these Arabic writers – if at all, or in what form – is unknown.

Though the Greek treatise by Galen is not preserved today, there are a considerable number of scattered references to eye diseases found throughout those treatises by Galen that are considered genuine and are preserved today. A search through the discussions of eye complaints in both the genuine and spurious works of Galen that are found in the printed texts available in the *Corpus Medicorum Graecorum* and in the edition by C. G. Kühn, as well as the fragments of six papyri containing Galenic texts dating from the third to sixth century recently edited and studied by Marie-Hélène Marganne,[39] produced a general agreement as to the eye conditions mentioned (with one important exception to be discussed below), but presented in totally different contexts.

The Greek treatise preserved today that initially seemed the most likely candidate was the pseudo-Galenic *Introduction*, of which the sixteenth chapter is concerned solely with eye

34 For this catalogue of Galenic writings in Ya'qūbī's *Ta'rīkh*, see *Ibn-Wādhih qui dicitur al-Ja'qūbī, Historiae*, ed. by M. Th. Houtsma (Leiden 1883) I, 130_3-133_{bottom}. See also Martin Klamroth, 'Ueber die Auszüge aus griechischen Schriftstellern bei al-Ja'qūbī', *ZDMG 40* (1886) 614-34.

35 He might also have been using an informant who told him of what Syriac treatises were available. I thank Peter Starr for this suggestion.

36 Ibn Abī Uṣaybi'ah, *'Uyūn al-anbā'* (above, n. 9) II, 95_{29}-96.

37 A.I. Sabra, 'One Ibn al-Haytham or Two? An exercise in reading the bio-bibliographical sources', *Zeitschrift für Geschichte der Arabisch-Islamischen Wissenschaften 12* (1998) 1-40.

38 Ibn Abī Uṣaybi'ah, *'Uyūn al-anbā'* (above, n. 9) I, 90_{21}-103_7 presents a list of Galen's writings.

39 Marie-Hélène Marganne, *L'ophtalmologie dans l'égypte gréco-romaine d'après les papyrus littéraires grecs* (Leiden 1994).

diseases.[40] It also is essentially an enumeration of eye diseases with short statements of symptoms, but there the similarity ends. In the Greek *Introduction* some therapies are occasionally given, but never in the Arabic treatise; there is no discussion of ocular anatomy; and 104 diseases are enumerated, rather than 91 as found in the Arabic *jawāmi'*. Moreover, the diseases in the *Introduction* are grouped in a manner different from those in the Arabic treatise; for example, in the *Introduction* there is one group of diseases affecting the interior of the eyelid and another group affecting the exterior of the lids, while in the Arabic *jawāmi'* the conditions are combined under the general rubric 'diseases of the eyelid'.

There is an important general discrepancy between the Arabic *jawāmi'* of Galen's treatise on diseases occurring in the eye and all the fragmentary discussions of eye diseases in the genuine works of Galen, as well as the extended discussion in the spurious *Introduction* – and that is that this Arabic treatise includes in its enumeration of diseases of the conjunctiva the condition called in Arabic *al-sabal*, meaning 'the rain'.

There is no conclusive evidence at this point that Galen, or any physician prior to the ninth century, recognized this particular ocular disorder. Certainly in the treatises by Galen now preserved there appears to be no description of the condition that in Arabic is called *al-sabal* and that today we call *trachomatous pannus*. Pannus (Latin meaning 'a piece of cloth') is an invasion of the cornea by vessels from the limbus, and occasionally the entire cornea becomes vascularized. It was classified by medieval Islamic physicians as a disease of the conjunctiva. Yūḥannā ibn Māsawayh (who died in 857 and was the teacher of Ḥunayn) designated it a sequela of trachoma and described a procedure for removing it, using tiny hooks and a scalpel. Ibn Māsawayh used for the condition not only the Arabic names *al-sabal* ('rain') and *rīḥ al-sabal* 'the pouring of rain' but also a Persian term *bārandū'*, apparently a misspelling of *bārandagī* which also means 'pouring of rain'.

Ḥunayn in his 'Ten Treatises on the Eye' included the condition, using the name *al-sabal*. He did not give the Persian name, but rather stated that the Greek word for the condition was *qīrsūfthālmiyā*, an Arabic transliteration apparently for a Greek word *kirsophthalmia*, from *kirsos* meaning an enlargement of a blood vessel and *ophthalmia*, a disease of the eye.[41] The Greek word, however, is not known to occur in any Galenic treatise or any other Greek medical writing. The Greek term would, of course, fit well with the vascularization of the cornea, which was the hallmark of trachomatous pannus. It is significant that in this Arabic summary of Galen's treatise there is no transliterated Greek term given in this particular instance. Very often, but not always, after giving the Arabic term, the author of the summary then states what it would be called in Greek, *bi-l-yūnānīyah*, transliterating the word.

If we could demonstrate that Galen did actually know of this condition, then we might be more inclined to view the summary as reflecting a genuine work of Galen. An additional piece of evidence supporting the notion that Greek physicians, such as Galen, recognized the condition comes from al-Rāzī (d. 925), for in the *Ḥāwī* he has three passages from Galen which described eye conditions given the name *al-sabal*. However, I have argued elsewhere that these passages cannot be used as conclusive evidence of Galen's knowledge of the

40 Galen 14, 767-77 K.

41 Max Meyerhof, *The Book of the Ten Treatises on the Eye ascribed to Hunain ibn Ishaq (809-877 AD): The earliest existing Systematic Text-book of Ophthalmology* (Cairo 1928) 57 (Arabic text, p. 130).

condition, since the symptoms actually described are not those of *pannus* and since his use of the term *sabal* can be explained by Rāzī's misreading another word written in Arabic.[42]

Perhaps a more convincing piece of evidence that Galen knew of this condition is to be found in Galen's treatise *On Examinations by Which the Best Physicians are Recognized*. This treatise is preserved only in its Arabic translation, recently edited by Iskandar. There we read, in an anti-surgical polemic: 'You should praise a physician's knowledge of drugs if he can cure...without resorting to incision... You should also praise anyone who cures, with drugs alone, diseases of the eye which others treat by excision'.[43] Galen then lists ten ocular conditions usually treated by surgery, and the list includes pannus (*sabal*), in addition to pterygium, trachoma, chalazion, cataract and other operable disorders. However, the crucial word does not occur in the earlier of the only two extant manuscripts (copied in 1572), or in the extensive excerpt of this passage given by Ibn Abī Uṣaybi'ah, but only in the eighteenth century manuscript. Furthermore, in a later passage where operable eye conditions are again ennumerated, *sabal* (pannus) is found to be missing.[44]

Therefore, our best evidence is that Ḥunayn stated that the Greeks knew of the condition we call *pannus* and that they called it *kirsophthalmia*, and that an eighteenth-century copy of a Galenic treatise has the word when earlier copies do not. It is of course entirely possible that it was described and perhaps treated by Hellenistic and Byzantine physicians but not mentioned in the preserved Greek literature. But at this point we cannot know with certainty whether Galen knew of the condition or not. And on the unresolved question of whether or not Galen knew of the condition called *sabal* hangs our ability today to state with certainty whether this is a summary of a genuine work of Galen.

There are some other features of the Arabic text that might provide additional clues as to its origin. One of them is the nature of the transliterated Greek words – for in many instances they appear truncated or wildly off the mark. It is possible they represent a transliteration from another language, such as Syriac, or perhaps the manuscript tradition itself is corrupt.

Another important feature is that the medical vocabulary and some of the ideas and classifications are much closer to those found in a treatise by Ibn Māsawayh than to any subsequent ones that I have located, including those by Ḥunayn. Ibn Māsawayh's small treatise *Kitāb Daghal al-'ayn* ('The Unsoundness of the Eye') is the earliest preserved monograph on the eye, and curiously the only two manuscript copies are in the St Petersburg and Cairo volumes that also have the *jawāmi'* of Galen's treatise on eye diseases. It is in this small treatise, *Kitāb Daghal al-'ayn*, that Ibn Māsawayh gave the Persian term for the condition usually called *sabal*. Although the Persian term does not occur in the *jawāmi'*, in other ways the terminology (some of it rather distinctive) and some concepts in *Kitāb Daghal al-'ayn* are strikingly similar to those found in this *jawāmi'*.

Another notable feature of the *jawāmi'* is that the starting point is an Arabic medical vocabulary that is then defined in terms of Greek. This might suggest that it is not a translation from the Greek. A counter-argument to that, however, would be that Ḥunayn in some of his

42 See E. Savage-Smith, 'Hellenistic and Byzantine Ophthalmology: trachoma and sequelae', *DOP 38* (1984) 169-86, esp. 170-72, for a discussion of these passages.

43 A. Z. Iskandar, *Galen On Examinations by which the best physicians are recognized*, *CMG*, Suppl. Or. IV, (Berlin 1980) 117.

44 *Ibid* 137.

translations of Galenic treatises first translated a sentence or phrase or word into Arabic and then would insert the expression, and 'it is known in Greek as ...'.

Could this *jawāmi'* then be an extract of Galen's lost *On the Diagnosis of Eye Diseases*, set into a late-antique didactic format of ennumeration and branch-diagramming? Or, was it an early Arabic composition employing a format associated (at least in the ninth and tenth-centuries) with the prestigious Alexandrian school and drawing upon the authority of Galen.

The attribution in many of the manuscripts containing various Galenic *jawāmi'* to the Alexandrians indicates that in the minds of ninth- and tenth-century Islamic physicians Alexandria was associated with the teaching of Galenic medicine, and that for a manuscript to suggest an association with Alexandria was to enhance its authority and possibly authenticity. It also indicates that Alexandria at that time had a reputation for producing summaries of treatises, even though Ḥunayn makes no mention of such summaries. It also implies that Alexandria had a reputation for distinctive didactic methods of presentation, such as tabular presentation or branch-diagramming or possibly question-and-answer. But there is no secure evidence that such techniques actually were a part of the Alexandrian scene.

Clearly the short classification of diseases and basic symptoms suggests a catechism for learning names and categories of eye diseases, and the branch-diagrams employed in some manuscripts support the didactic and mnemonic value of this tract and others which employ such techniques.

There are additional questions that remain unsettled: Who was producing such summaries, and for what community? Presumably an Arabic-speaking student wanting a short-cut to memorizing eye diseases and their symptoms. However, why would the Greek terms be included? Does that mean that knowledge of Greek terminology was expected of the learned Arabic-speaking physician when this tract was composed? And was this *jawāmi'* composed for the same audience, or type of audience – possibly bi-lingual – who made use of the other summaries, or *jawāmi'*, of Galenic treatises that circulated in Syriac (according to Ḥunayn) and in Arabic in the ninth century, both having possibly been translated from the Greek?

I am inclined to think that the *Jawāmi' Kitāb Jālīnūs fī al-amrāḍ al-hādith fī al-'ayn* may well represent an abbreviated version of Galen's own treatise, although we cannot rule out the possibility that it was an early Arabic composition. I have been slow to come to this position – not least because I've gone into print saying that Galen probably did not know of *pannus* – but perhaps I was wrong, and we do have here a condensed version, in Arabic translation, of Galen's own treatise *On the Diagnosis of Diseases that Occur in the Eye*.

The Oriental Institute, University of Oxford

THE RECEPTION OF GALEN IN MAIMONIDES' *MEDICAL APHORISMS*

GERRIT BOS

Abū 'Imrān Mūsā ibn 'Ubayd Allāh, usually called Moses Maimonides, one of the greatest philosophers and experts in Jewish law (Halakhah),[1] was an eminent physician as well. Born in Cordoba in 1138,[2] he was forced to leave his native city at the age of thirteen because of persecution by the fanatical Muslim sect of the Almohads and the policy of religious intolerance adopted by them. After a stay of about twelve years in southern Spain, the family moved to Fez in the Maghreb. Some years later, probably around 1165, they moved again because of the persecutions of the Jews in the Maghreb, this time to Palestine. After some months the family moved on to Egypt, and settled in Fusṭāṭ, the ancient part of Cairo. It was here that Maimonides started to practise and teach medicine alongside his commercial activities in the India trade.[3] Thus he became the physician of al-Qāḍī al-Fāḍil, the famous counsellor and secretary to Saladin.[4] Later on he became court physician of al-Malik al-Afḍal, after the latter's accession to the throne in the winter of 1198-9. It is generally assumed that Maimonides died in 1204. The theory that for some years he served as *Nagid* (Head) of the Jewish community seems to be unfounded.[5]

1 For Maimonides' biographical data see E. J. Vajda, 'Ibn Maymūn', *The Encyclopaedia of Islam*, new ed. (Leiden and London 1960-) (= E.I.^{2}), vol. 3,. 876-88; L. I. Rabinowitz,. 'Maimonides, Moses' *Encyclopaedia Judaica*, 16 vols (Jerusalem 1971) vol. 11, 754-64; Bernard Lewis, 'Maimonides, Lionheart and Saladin', *Eretz-Israel* 7 (1964) 70-75; S. D. Goitein, 'Ḥayyei ha-Rambam le-Or Gilluyim ḥadashim min ha-Genizah ha-qahirit', *Perakim* 4 (1966) 29-42; idem, 'Moses Maimonides, man of action, a revision of the master's biography in light of the Geniza documents', in *Hommage à Georges Vajda: Études d'histoire et de pensée juives*, Louvain 1980, pp. 155-167; Isaac Shailat, *Iggerot ha-Rambam,* 2 vols (Jerusalem 1987-8) 1, 19-21; Mark R. Cohen, 'Maimonides' Egypt', in *Moses Maimonides and his time*, ed. Eric L. Ormsby (Washington 1989) 21-34; Menahem Ben-Sasson, 'Maimonides in Egypt: the first stage,' in *Maimonidean studies*, ed. Arthur Hyman, vol. 2 (New York 1991) 3-30; Jacob Levinger, 'Was Maimonides "Rais al-Yahud" in Egypt?', in *Studies in Maimonides*, ed. Isadore Twerski (Cambridge, Mass. And London 1990) 83-93.

2 While traditionally his date of birth is set at 1135, Maimonides himself wrote in 1168, in the colophon to his *Commentary on the Mishnah,* that he was then in Egypt and thirty years old; Goitein, *Moses Maimonides, man of action*, above, note 1, 155, argued on the basis of this that the actual year of his birth should be put at 1138; see also J.O. Leibowitz, 'Maimonides: Der Mann und sein Werk. Formen der Weisheit', *Ariel* 40 (1976) 73-78 (75-76).

3 Goitein, *Moses Maimonides*, above, note 1, 163, showed that Maimonides was already involved in this trade before his younger brother David perished in a shipwreck in 1169, and that he still had a hand in it in 1191 when he was practising as a physician.

4 See C. Brockelmann-[Cl. Cahen]), 'al-Kāḍī-al-Fāḍil', in E.I.^{2}, vol. 4, 376-77.

5 See Herbert A. Davidson, 'Maimonides' putative position as official head of the Egyptian Jewish community', in *Ḥazon Naḥum. Meḥkarim ba-Halakhah, Maḥshavah we-Historyah Yehudit muggashim le-Dr. Le'om be-haggi'o le-sevah*, ed. J. Elman and S. Gurag (New York 1998) 115-28.

Maimonides' Medical Aphorisms

Maimonides was a prolific author in the field of medicine, composing ten works considered as authentic.[6] The most famous and most voluminous is without any doubt the *Medical Aphorisms*. The date of its composition is uncertain. On the one hand, we have indications that it belongs to the earlier medical works composed by Maimonides,[7] since it is quoted by him in his *Commentary on the Aphorisms of Hippocrates*[8] and since this latter work is mentioned in turn by him in his *Treatise on Asthma.*[9] On the other hand, a scribal note to the Arabic text extant in MS Gotha 1937[10] seems to show that at least the twenty-fifth treatise was composed at the end of his life. According to this note, Maimonides' nephew Abū l-Ma'ālī b.Yūsuf b. 'Abdallāh copied the text of the first twenty-four *Maqāla*s once it had been arranged and put into order by Maimonides in his presence; the twenty-fifth *Maqāla,* however, was arranged and copied by him in the beginning of the year 602 A.H. (= August 1205) after the death of Maimonides. From this note some scholars have concluded that the entire extant version of the *Medical Aphorisms* should be considered as his last work.[11] Recently, Lieber has suggested that the first twenty-four treatises were indeed composed at an early date, and that the twenty-fifth treatise was composed at the end of his life and then added to the earlier treatises.[12]

The *Medical Aphorisms* enjoyed great popularity in medieval western Europe. In the thirteenth century it was translated into Latin, and at least two different versions are extant in many editions. The translation by John of Capua was printed in Bologna in 1489 and in Venice in 1497 and was rapidly followed by numerous other editions.[13] From the thirteenth until the fifteenth centuries it was 'the most widely known and wanted repetitorium of Galen',[14] and was quoted and recommended as a medical textbook by Jean de Tournemire, professor of medicine in Montpellier (d. 1396), who called it *Flores Galieni.* In Jewish circles it became influential and popular through the two Hebrew translations, one prepared by Nathan ha-Me'ati between 1279 and 1283 in Rome[15] and the second prepared in 1277 by

6 For a detailed survey of his medical works see the survey in Gerrit Bos, Maimonides, *Treatise on asthma* (Provo, Utah 2002).

7 According to M. Ullmann, *Die Medizin im Islam* (Leiden, Cologne 1970), it was probably composed between 1187 and 1190.

8 II, 33, following Moses Ibn Tibbon's Hebrew translation as edited by S. Muntner, *Rabbeinu Mosheh Ben Maimon. Perush le-Phirkei Abuqraṭ* (Jerusalem 1961); English trans.: F. Rosner, *Maimonides' Commentary on the Aphorisms of Hippocrates,* Maimonides' Medical Writings 2 (Haifa 1987).

9 Ed. Gerrit Bos (Provo, Utah 2002): XIII, 19.

10 For this manuscript cf. W. Pertsch, *Die arabischen Handschriften der herzoglichen Bibliothek zu Gotha* (1883).

11 According to Max Meyerhof, 'The medical work of Maimonides', in *Essays on Maimonides, an octocentennial volume*, ed. S. W. Baron (New York 1941) 265-99 (276), it was composed during the last ten years of his life; J.O. Leibowitz, 'Maimonides' Aphorisms', *Koroth* 1 (1955) I, remarks that it was rewritten at the end of his life.

12 Cf. Elinor Lieber, 'Maimonides the medical humanist', in *Maimonidean Studies*, vol. 4, ed. A. Hyman (New York 2000) 51.

13 J. O. Leibowitz, 'The Latin translations of Maimonides' Aphorisms', *Koroth* 6 (1970) 273-81.

14 S. Muntner, *Rabbeinu Mosheh Ben Maimon. Hanhagat ha-Beri'ut* (Jerusalem 1957) XIII.

15 Nathan ha-Me'ati's translation was edited for the first time in Lemberg in 1834 and reprinted in Vilna in 1888. A new edition was prepared by S. Muntner, *Rabbeinu Mosheh Ben Maimon. Pirkei Mosheh ba-refu'ah* (Jerusalem 1959). This edition was the basis for the translation by F. Rosner and S. Muntner, *The Medical Aphorisms of Moses*

Zeraḥyah Ben Isaac Ben She'altiel Ḥen, equally of Rome.[16] The famous Italian Jewish philosopher Hillel Ben Samuel of Verona (ca. 1220-1295) wrote to his friend Isaac, known as Master Gaio, the pope's doctor, requesting the Hebrew translations by Nathan ha-Me'ati of both the *Medical Aphorisms* and the *Commentary on Hippocrates' Aphorisms*.[17] Hillel writes to Isaac that he should not care about the cost but make a scribe copy them, whatever the price, since he loves these books so much. In an anonymous compilation on fevers written in old French in Hebrew script and probably dating from the 13th century the author remarks that he wants to translate everything from the 'Pirkei Rabbenu Mosheh' concerning the crisis <of fevers>.[18] And a recipe for quartan fever featuring in *Medical Aphorisms* X, 56 and derived from Galen's *De theriaca ad Pisonem* is quoted by Salmias (Salamias) of Lunel in a treatise on fevers extant in the Bodleian library in Oxford in the name of Maimonides.[19] Gerson Ben Ezechias of Beaumes (born around 1373), the author of a medical handbook in rhyme, remarks that he has read in the presence of the nasi Todros of Narbonne or his son Kalonymos a variety of medical treatises, amongst which the 'Chapters of Moses'.[20] A gloss in the *Sefer ha-kibbusin*, a treatise on purgatives translated from the Latin gives a recipe for headaches derived from the *Pirkei Mosheh*.[21] Rabbi Solomon Ben Abraham ha-Kohen, who lived in 16th century Greece, quotes in his *Responsa* Aphorism XVI, 18 about a woman who suffered from hysterical suffocation and who through touching her genitals when she inserted medicines into her womb had an experience similar to an orgasm and then discharged a thick clot of semen and thus found relief from her afflictions.[22] The popularity of the *Medical Aphorisms* in Jewish circles confirms Conrad's suggestion that 'the Greek medical corpus was mediated to physicians of the Jewish communities primarily as extracts and easily

Maimonides, 2 vols. (New York 1970-71), and the revised translation by Rosner, *The Medical Aphorisms of* Moses *Maimonides,* Maimonides' Medical Writings 3 (Haifa 1989). For Nathan ha-Me'ati (of Cento), see H. Vogelstein and P. Rieger, *Geschichte der Juden in Rom* (Berlin 1896) vol. 1, 399-402; M. Steinschneider, *Die hebräischen Übersetzungen des Mittelalters und die Juden als Dolmetscher* (Berlin 1893) 766.

16 On Zeraḥyah see Vogelstein-Rieger, *Geschichte,* above, note 15, vol. 1, 271-275, 409-418; A. Ravitzky, *Mishnato shel R. Zeraḥyah b. Isaac b. She'altiel Ḥen we-ha-Hagut ha-Maimunit-Tibbonit ba-Me'ah ha-Shelosh-Esreh*, Ph. D. Diss. (Jerusalem 1977) 69-75; G. Bos, *Aristotle's De Anima. Translated into Hebrew by Zeraḥyah b. Isaac b. She'altiel Ḥen* (Leiden 1994) ch. 7: 'Zeraḥyah's technique of translation.' For the manuscripts of this translation see Steinschneider, *Die hebräischen Übersetzungen*, above, note 15, 766.

17 See B. Richler, 'Another Letter from Hillel Ben Samuel to Isaac the Physician?', *Kiryat Sefer* 62 (1988-89) 450-52; Joseph Shatzmiller, *Jews, Medicine and Medieval Society* (Berkeley and Los Angeles 1994) 46.

18 See Steinschneider, *Verzeichnis der hebräischen Handschriften in Berlin*, 2 vols (Berlin 1878-97: repr. in 1 vol., Hildesheim 1980) no. 233; idem, 'Eine altfranzösische Compilation eines Juden über die Fieber', *Virchows Archiv* 127 (1894) 401; cf. Gerrit Bos, ed. *Medical Aphorisms* I-V, introduction (forthcoming).

19 Cf. A.. Neubauer, *Catalogue of the Hebrew Manuscripts in the Bodleian Library* (Oxford 1886, repr. 1994) , no. 2133, fols 45a-45b And: *Supplement of Addenda and Corrigenda* compiled under the direction of Malachi Beit-Arié and edited by R.A. May (Oxford 1994), cf. Henri Gross, *Gallia Judaica. Dictionnaire géographique de la France d'après les sources Rabbiniques.* Traduit sur le manuscrit de l'auteur par Moïse Bloch (Paris 1897) 288-289, no. 25. Salmias quotes from the Hebrew translation composed by Nathan ha-Me'ati.

20 Cf. E. Renan, *Les écrivains juifs français du XIVe siècle* (Paris 1893: repr. Farnborough 1969) 438.

21 Cf. Steinschneider, *Verzeichnis der hebräischen Handschriften in Berlin*, MS no. 113 (p. 93), fol. 35a.

22 *Shut Maharshakh* II, 105.

learned statements, rather than through the fundamental works where medical theory and practice were worked out in detail.'[23]

Medical Aphorisms consists of 25 treatises, each dealing with a subspecialty in medicine: 1-3: anatomy, physiology and general pathology; 4-6: symptology; 7-14: aetiology; 15: surgery; 16: gynaecology; 17: hygiene; 18-20: diet; 21-22: pharmacopoeia; 23: explanation of obscure names and conceptions in Galen's works; 24: collection of rare and interesting cases out of Galen's writings; 25: criticism of Galen.[24] Of this work the following sections have been published so far in the original Arabic: 1. The introduction and six fragments.[25] 2. The section in treatise XXV which contains Maimonides' criticism of the philosophical doctrines adhered to by Galen,[26] 3. Treatise XXV, 56-71.[27] The complete text of the *Medical Aphorisms* is only available in an edition of the medieval Hebrew translation prepared by Nathan ha-Me'ati[28] and in modern English translations.[29] The Hebrew edition of the *Medical Aphorisms* as prepared by Muntner is unsatisfactory according to our modern standards of editorship, which demand a critical edition of the text, based on an evaluation of all the available MSS and a critical apparatus referring to deviant readings. Moreover, this edition suffers from many editorial mistakes, omissions and additions. The modern English translations by Rosner-Muntner and Rosner are flawed not only because they are based on corrupt editions, but also because of the many mistakes and misunderstandings of the Hebrew text. Thus they are unreliable and cannot be considered to represent the true words of the author. Therefore, a publication of a critical edition with translation of the Arabic original of the *Medical Aphorisms* is an urgent desideratum. A project was set up in 1995, initially at University College London sponsored by the Wellcome Trust in London and now at the University of Cologne supported by the Deutsche Forschungsgemeinschaft, to edit the *Medical Aphorisms* and the other unpublished medical works composed by Maimonides. These were his *Treatise on Asthma* (now in press); his *Commentary on Hippocrates' Aphorisms*; his *Treatise on Poisons,* and *Extracts from Galen.* At a later stage we hope to look closely at those works which have appeared in critical editions in the original Arabic in the past but possibly have to be revised especially in the light of new manuscript findings. These works are: *Treatise on Hemorrhoids*, *Treatise on Cohabitation*, and *Regimen of*

23 See Review of Haskell D. Isaacs, *Medical and paramedical manuscripts in the Cambridge Genizah Collections*, *Bulletin of the School of Oriental and African Studies* 59 (1996), 136-37.

24 See J. Schacht, M. Meyerhof, 'Maimonides against Galen, on Philosophy and Cosmogony,' *Bulletin of the Faculty of Arts of the University of Egypt* 5 (1937) Arabic Section. 53-88 (59-60). For a general introduction to the *Medical Aphorisms* see Muntner, *Pirkei Mosheh ba-Refu'ah,* pp. XII-XVIII; F. Rosner, 'The Medical Aphorisms of Maimonides', *Muntner Memorial Volume* (Jerusalem 1983) 6-30; idem, *The Medical Aphorisms of Moses Maimonides*, above, note 15, XII-XXVI.

25 P. Kahle, 'Mosis Maimonidis Aphorismum Praefatio et Excerpta,' in *Galeni in Platonis Timaeum Commentarii Fragmenta,* ed. H. O. Schröder, *CMG* Suppl. I, Appendix 2.

26 Schacht-Meyerhof, *Maimonides against Galen*, above, note 24.

27 See Y. Kafiḥ, *R. Moshe Ben Maimon, Iggarot, Nispaḥ* II (Jerusalem 1987) 148-67. Kafiḥ transcribed the original Arabic text from MS Gotha into Judeo-Arabic.

28 Ed. Muntner (Jerusalem 1959), above, note 15.

29 Rosner-Muntner, *The Medical Aphorisms of Moses Maimonides*, and the revised edition by Rosner, *The Medical Aphorisms of Moses Maimonides,* above, note 15.

Health, all of them edited by Kroner. The series is to be published in the Islamic Science Series edited by Glen M. Cooper, of Brigham Young University.

Most of the aphorisms featuring in the *Medical Aphorisms* are based on the works of Galen, some of them going back to Galenic works which are no longer available in the original Greek.[30] An example is Aphorism VII, 71:[31] 'As for laughter occurring through the tickling of the armpits and the soles of the feet as well as laughter which occurs when seeing or hearing comical things, it is absolutely impossible to find out its cause.' This aphorism is a quotation from Galen's *De motibus dubiis* and only survives in two different Latin translations by Marcus of Toledo and Niccolò da Reggio[32] and in an Arabic translation entitled *Fī l-ḥarakāt al-mu'tāṣa* which has been preserved in Istanbul, MS Ayasofya 3631 and which I am currently editing. Muntner assumed that the Hebrew translation of Galen's *De motibus dubiis*, entitled *Ba-Tenu'ot ha-Mukhraḥot*, was identical with Galen's *On the movements of the chest and lung, De motibus thoracis et pulmonis.*[33] This assumption was adopted by Rosner as well.[34] On another occasion Maimonides quotes from a text under a title that has not been preserved in bibliographical literature, to the best of my knowledge.[35] Maimonides remarks (IX, 101):

> Much flesh and fat is harmful and detrimental and gives the body an ugly appearance and hinders its activities and movements. Therefore, those suffering from it should travel a lot and do much walking in the sun. Travel over sea is especially good because the sea air dissolves the <superfluous> moistures. One should feed them foods with little nourishment such as vegetables and that which contains heat such as onions, garlic and salted fish and that which strengthens but does not moisten such as meat roasted from non-fat meats. One should not let them bathe in hot water except that of thermal springs. One should keep them a little thirsty and make their bodies firm in any possible manner. *De extenuatione corporum pinguium.*

Although the Arabic title of this aphorism, namely *Fī tahzīl al-ajsād al-'abla* (A slimming <diet> for fat bodies) is reminiscent of Galen's *Fī l-tadbīr al-mulaṭṭif* (*On the thinning diet, De diaeta subtiliante*) it is unlikely that it is just another name for it, especially because this

30 The following analysis of the *Medical Aphorisms* is based on my research into Books I-XXI.

31 Numeration according to the Hebrew edition by Muntner, *Pirkei Mosheh bi-Refu'ah* and the English translation in Rosner, *Medical Aphorisms*, above, note 15.

32 Cf. Carlos J. Larrain, 'Galen, De motibus dubiis: Die lateinische Übersetzung des Niccolò da Reggio', *Traditio* 49 (1994) 171-233 (223). See above, pp. 79-85.

33 See S. Muntner, 'Zihhuy Sifrei Galenus ha-Nizkarim 'al Yedei ha-Rambam be-Sifro 'Pirkei Mosheh'' (Identification of the works by Galen mentioned by Maimonides in his Medical Aphorisms), *Ha-Rofe ha-Ivri* 27 (1954) 120-33 (132), and *Pirkei Mosheh ba-refu'ah*, a.l.

34 Cf. *The Medical Aphorisms of Maimonides*, a.l. For another quotation from *De motibus dubiis* see Aphorism IX, 11.

35 Cf. Galen, *De libr. propr.* XIX 8-48 K.; G. Bergsträsser, *Ḥunayn ibn Isḥāq über die syrischen und arabischen Galen-Übersetzungen,* Abh. für die Kunde des Morgenlandes 17,2 (Leipzig 1925); idem, *Neue Materialien zu Ḥunayn ibn Isḥāq's Galen-Bibliographie,* Abh. für die Kunde des Morgenlandes 19,2 (Leipzig 1932); Moritz Steinschneider, 'Die griechischen Aerzte in arabischen Übersetzungen', *Virchows Archiv* 124 (1891) 115-36, 268-96, 455-87; Max Meyerhof, 'Über echte und unechte Schriften Galens, nach arabischen Quellen', *SBAW* 28 (1928) 533-48; G. Strohmaier, 'Der syrische und arabische Galen', *ANRW* II:37.2 (1994) 1987-2017.

quotation does not feature in the text edited by Kalbfleisch.[36] Moreover, the difference in title between the two Galenic treatises is maintained by both Hebrew translators. The title *Fī tahzīl al-ajsād al-'abla* is translated by Nathan ha-Me'ati as *Be-hattakhat ha-geshamim he-'avim we-ha-shemenim* and by Zeraḥyah Ben Isaac Ben She'altiel Ḥen as *Be-harazut(!) ha-guf ha-shamen,* while *Fī l-tadbīr al-mulaṭṭif* is translated by both as *Ba-hanhagah ha-medaqdeqet.* On another occasion Maimonides quotes from a treatise bearing a title different from that under which it is known in the bibliographical literature. This is the case with his quotation from the pseudo-Galenic treatise *De theriaca ad Pamphilianum,* which is also the first to be documented in the Arabic medical tradition.[37] Maimonides quotes this treatise as *The Treatise on the Preparation of the Theriac* (*Maqāla fī 'amal al-tiryāq*) although it is usually referred to as *K. al-tiryāq ilā Bamfūliyānūs*.[38]

Sometimes Maimonides quotes from treatises ascribed to Galen whose authenticity is doubtful. This is the case for his quotations from *De signis mortis* (*On the signs of death*). An example is the following aphorism (III, 76):

> Chronic illnesses such as orthopnea, stones, tumours in the nose, bad ulcers and the like, most of these afflictions which occur to youngsters and children are cured in forty days or seven months or seven years. Some of them <are cured> before the pubic hair has started to grow and in the case of girls until the time of menstruation. *De signis mortis.*[39]

This work does not feature in the lists of works composed by Galen or ascribed to him.[40] The Arabic tradition has preserved a number of texts under this title ascribed to Hippocrates, all in manuscript, whose central subject is the signs of death derived from the efflorences of the skin. A text under this title ascribed to the famous philosopher and physician al-Kindī is actually an adaptation of a part of Hippocrates' *Prognostics*.[41] And sometimes Maimonides quotes from a pseudo-Galenic treatise that has only been preserved in an Arabic translation although the passage he adduces is missing from that particular translation. For example, at *Aphorisms* VII, 51 he quotes from Galen's *De somno et vigilia*:

> If someone wishes to raise his voice, he should get used to open his mouth widely so that he lets in much air which widens the larynx. For then the voice is loud. Therefore, those whose larynx is narrow and small have thin, small voices without any substance so that they are broken off quickly. But those whose larynx is wide have a full and powerful voice. Children, women and eunuchs have thin and weak voices because their larynx is small.

Since Maimonides' quotation does not feature in this Arabic translation, Nabielek, its editor and translator, concludes that the original text of this treatise must have been more

36 C. Kalbfleisch, Galen, *De victu attenuante, CMG* V 4, 2, 431-51. Cf. above, pp. 47-56.

37 Cf. *Medical Aphorisms* XX, 52.

38 Cf. Ullmann, *Die Medizin im Islam*, above, note 7, 49, no. 52.

39 For another quotation from this work see *Aphorisms* VI, 48.

40 See above, note 35.

41 See Gerrit Bos, 'A recovered Fragment on the Signs of Death from Abū Yūsuf al-Kindī's "Medical Summaries"', *Zeitschrift für Geschichte der arabisch-islamischen Wissenschaften* 6 (1990) 189-94.

extensive.[42] At the same time he concludes paradoxically that because of the similarity between Maimonides' text and the relevant passage from Oribasius' *Collectiones Medicae* the latter must have been the source he copied from.[43]

On many occasions Maimonides quotes from pseudo-Galen's commentary on Hippocrates' *De humoribus.*[44] These quotations are a valuable source for reconstructing part of Galen's genuine commentary to this book since the text edited by Kühn is, as Deichgräber showed, a Renaissance forgery.[45] On that occasion Deichgräber also demonstrated that Maimonides' quotations closely parallel those of Oribasius. But at the same time he warned against over-optimism since both Maimonides and Oribasius tend to abbreviate the original Galenic text.[46] Deichgräber's analysis of these quotations also shows the importance of the availability of a critical original text of Maimonides' *Medical Aphorisms.* For in one case his conclusion that Maimonides' summary of the Galenic text is unique for its brevity insofar as it summarises 15 lines of the Greek text in one line is based on a corrupt Latin version.[47]

Maimonides quotes not only from authentic and inauthentic Galenic works, but also from other ancient and medieval physicians. Thus, he quotes from a lost commentary by a certain Asclepius on Hippocrates' treatise *On fractures*:

> <Broken> bones of young people heal sooner than those of children, because children need material for their growth and to replace the matter which has been dissolved. This was mentioned by Asclepius in the first treatise of his commentary on <fractures> and their setting.[48] Do not attempt to set any broken bone if only four days have passed, lest you cause the patient severe harm. The third <treatise> of Asclepius' commentary on <Hippocrates'> book on <fractures> and their setting.[49]

This commentary by Asclepius is not mentioned in the bibliographical literature. Leclerc refers to a commentator called *Senflious* on Hippocrates' treatise *On fractures* quoted in al-Rāzī's *K. al-Ḥāwī*, and remarks that the Latin translator has rendered the name of this commentator as *Herilius* or *Sterilius*. He adds that this name could also be read as Simplicius.[50] Steinschneider[51] further notes that the Arabic term used (*Sinblikius*) is indeed the first commentator on Hippocrates mentioned by Ibn al-Nadīm in his *K. al-fihrist*,[52] although this name does not appear in the list of commentators given by Littré.[53] Ullmann identifies the

42 R. Nabielek, *Die ps.-galenische Schrift 'Über Schlaf und Wachsein' zum ersten Male herausgegeben, übersetzt und erläutert,* Diss. (Berlin 1977) 29; for another otherwise lost quotation from this treatise see *Aphorisms* XVII, 4.

43 See the comparative table, ibid., 33-34.

44 Cf. *Aphorisms* VIII, 55, XII, 43; XIII, 20, 29, 37; XIV, 3.

45 K. Deichgräber, 'Hippokrates' De humoribus in der Geschichte der griechischen Medizin' *AAWMainz*. 1972,. Nr. 14, p. 43.

46 Ibid., 45.

47 Cf. ibid., 51 ; *Aphorisms* VIII, 55.

48 *Aphorisms* XV, 63.

49 *Aphorisms* XV, 64; for yet another quotation see XXI, 33.

50 L. Leclerc, *Histoire de la médecine arabe*, 2 vols (Paris 1876) 1, 235, 267.

51 M. Steinschneider, *Die arabischen Übersetzungen aus dem Griechischen* (repr. Graz 1960) 308.

52 *K. al-fihrist*, ed. Cairo, Miṭba'a al-istiqāma (1928) 415 (= 288,ed. Flügel).

53 *Oeuvres complètes d'Hippocrate*, 10 vols (Paris 1839-1861, repr. Amsterdam 1973-1989) 1, 81-132.

Arabic term used, i.e. the name of the commentator mentioned in al-Rāzī's *K. al-ḥawī*,[54] as Simplicius, acknowledging that he is otherwise unknown.[55] However, it seems more probable that the name of this commentator as quoted by al-Rāzī is a corruption of the Arabic (*'sqlbyws*) for Asclepius, and that he refers to the same Asclepius as Maimonides. Other quotations, called 'Rules in hortatory form' (*waṣāyā*) probably hail from a lost work entitled *Waṣīya*, that was composed by Abū l-'Alā' ibn Zuhr (d. 1131)[56] for his son Abū Marwān (d. 1162).[57] The central subject of these rules is treatment by means of purgatives. [58]Maimonides' comment on one of these rules give us a fine insight into both his critical attitude and his experience as a physician. For when Abū l-'Alā' remarks:

> It is a mistake to use musk as part of purgatives, and similarly to drink it with wine. Those who compound this remedy <and administer it> are mistaken, because they want to strengthen the organs and to let the medicine rise to the head, but forget that the effect of these purgatives is carried to the major organs, and sometimes such an organ cannot tolerate this and <the patient> is killed.[59]

Maimonides comments:

> This is correct if the purgation is done by poisonous drugs, such as pulp of colocynth or turbith (*Ipomoea turpethum*) because of their poisonous effect, or by strong drugs, such as laurel (*Laurus nobilis*) because of its strength. But safe drugs, and especially agaric which is good for poisons, are very beneficial if imbibed in wine. I have done so several times <and used such a drug> in order to cleanse the head and saw that it is very effective and that it cleanses the brain to a degree any <other> drug is incapable of. Moreover, the patient taking this drug found <new> energy and dilation of the soul. Therefore, consider the specific properties of the drugs which you administer.[60]

Another medieval physician extensively quoted by Maimonides is Ibn Wāfid (11th century) from Toledo. His pharmacopeia, which is available in a critical edition,[61] is Maimonides' primary source for a list of two hundred and sixty-five drugs to be applied internally and twenty drugs to be applied externally, which are common in all places and which every physician should know by heart.[62] Maimonides remarks that his decision to consult Ibn Wāfid's pharmacopoea was based on the fact that he is 'known for his skill and for his

54 al-Rāzī *K. al-Ḥāwī fī al-ṭibb*, vols. 1-23 (Hyderabad 1952-1974) 13, 159, although the spelling of the name in this edition is slightly different.

55 Ullmann, *Die Medizin im Islam*, above, note 7, 31, no. 12.

56 The title *Waṣīya* features in a quotation by Ibn al-Muṭrān (see Ullmann, ibid., 162); for Abū l-'Alā's literary activity see Ullmann, ibid., 162-63; Cristina Álvarez Millán, 'Actualización del corpus médico-literario de los Banõ Zuhr. Nota bibliográfica', *Al-Qanṭara. Revista de Estudios Árabes* 16 (1995) 173-80.

57 For Abū Marwān b. Zuhr (d. 1162), highly regarded by Maimonides, see Maimonides, *On Asthma* 9.1; and the bibliography in the preceding note.

58 See *Aphorisms* XIII, 44-49, 51.

59 *Aphorisms* XIII, 49.

60 *Aphorisms* XIII, 50.

61 *K. al-adwiya al-mufrada* (*Libro de los medicamentos simples*), ed. Luisa Fernanda Aguirre de Cárcer, 2 vols (Madrid 1995).

62 Cf. *Aphorisms* XXI, 68-87.

correct quotations from Galen and others'. Another medieval physician consulted and quoted by Maimonides for the pharmacological section of the *Medical Aphorisms* is al-Tamīmī, who hailed from Jerusalem and moved to Egypt in 970 to serve the vizir Ya'qūb ibn Killis. However, Maimonides' opinion about his pharmacological work entitled *K. al-murshid* (*The Guide*), which has only been preserved in part and is still in manuscript, was not so positive, as the following quotation shows clearly:

> This man who was on the Temple mountain and whose name is al-Tamīmī and who composed a book on drugs and called it *al-murshid*, allegedly had much experience. Although most of his statements are taken from others and although sometimes he wrongly understands the words of others, he still, in general, mentions many properties of various foods and of medications, and therefore I decided to write down those which were good in my opinion, whether foods or medicaments. [63]

Some of the Aphorisms quoted are in Galen's own words,[64] others are partly in Galen's words and partly in those of Maimonides, and yet other Aphorisms are completely reformulated by Maimonides, as he explicitly states in the introduction:

> I do not claim to have authored these aphorisms. I rather say that I have selected them, that is, I have collected them from what Galen said in his books, both what he said in his own treatises and what he said in his commentaries on Hippocrates. I have not been as fastidious with regard to these aphorisms as I was in my epitomes, where I quoted Galen's very words as I stipulated in the introduction to the epitomes. Instead, some of the aphorisms that I have selected are in the very words of Galen, or his and those of Hippocrates, since the two of them are combined in Galen's commentaries to Hippocrates' writings; for others the aphorism is partly Galen's words and partly my own; and for yet other aphorisms, it is my own words that express the idea that Galen mentioned.[65]

A constant element in the reformulation process is that of abbreviation. For instance, when speaking about the different symptoms of phrenitis Galen mentions the following: (*De locis affectis* V, 4 (VIII 330 K. transl. Siegel, 148-149): 'Sometimes we observe an irrational forgetfulness, when, for instance, the patients ask for the urinal but do not pass their water, or forget to surrender the urinal after having voided.' This text is summarized by Maimonides in *Aphorisms* VI, 37 as: 'forgetfulness of current things'. It is clear that Maimonides has summarized and abbreviated non-essential parts of this text. Another example of how Maimonides abbreviates Galen's discussion by omitting non-essential elements is the following quotation:

63 *Aphorisms* XX, 82.

64 These quotations are derived from the Arabic translations of Galen's works. For example, *Aphorisms* XVI, 5-7 are taken from Ḥunayn's translation of Galen's *De locis affectis* as featuring in MS Wellcome Or. 14a.

65 Y. Tzvi Langermann, 'Maimonides on the synochous fever', *Israel Oriental Studies* 13 (1993) 177. Steinschneider citing this introduction took issue with Leclerc's remark that the *Aphorisms* were literal quotations. See Steinschneider, *Die hebräischen Übersetzungen*, above, note 15, 765-66.

> Just as wine is extremely harmful for children, so for the elderly it is extremely beneficial. That wine is salutary for them which is the warmest and thinnest and which has a bright reddish colour. It is the one which Hippocrates calls 'tawny'. *De sanitate tuenda* V.[66]

Maimonides omits all the concrete examples which Galen adduces of those Greek and Italian wines having either one of these properties. Yet he is not satisfied with merely quoting or re-formulating the Galenic text. In several instances we see that Maimonides reflects upon these texts, and arrives at certain conclusions regarding their validity.Thus, he remarks that Galen's statement that the cause for laughing when the armpits are tickled or when one sees funny things cannot be known is correct because laughter is a specific characteristic of human beings and that the specific characteristics of any species of animals, plants and minerals cannot be explained aetiologically. In other cases Maimonides adds elements to the text, probably since these are essential in his view. For instance *Aphorism* III, 73 is taken from Galen, *De sanitate tuenda* VI, 5-6: 'If you find that someone falls ill only on rare occasions, do not change any of his habits in his whole way of life. But if someone falls ill frequently, you should look for its cause and eliminate it'. But Maimonides then adds: 'There is no doubt that this <occurs> through a change in one or more of his habits. You should also consider concerning someone whose habit you want to change whether or not such a change can be well tolerated by him. *De sanitate tuenda* I'.[67] In *Aphorism* I, 62 (= *De usu partium* IV, 18)[68] Maimonides presents the theory proposed by Galen that the right kidney lies higher than the left kidney, adding correctly that this holds good only for some living beings. May suggests that Galen was almost certainly describing conditions in some species of ape.[69] Maimonides' additions sometimes assume the form of an explicit comment, introduced by 'Says Moses' (*qāla Mūsā*), as is the case in XVI, 4:

> Says Moses: The aforementioned causes <of the retention of the menstrual blood> mentioned by him act only on the part of an organ, while menstrual blood is still present. He did not mention here those causes which are dependent on the blood, namely when its quantity greatly diminishes, for then there is no blood present any more. Therefore, he speaks of the retention of the menstrual blood and not of its cessation.

Some additions are of primary importance for our evaluation of Maimonides' theoretical and practical medical knowledge, especially in the field of pharmacology, a subject he was particularly interested in.[70] Thus, in *Aphorism* IX, 57 he comments on Galen's remark that 'if someone suffers from indigestion and the like and from a burning in the stomach which is so severe that one imagines that there is an inflamed tumor over there, he will benefit from a salve prepared with quince oil,'[71] that 'it is his practice to heat the oils in a bain-Marie'

66 *Aphorisms* XVII, 28.

67 *De sanitate tuenda* I, 5-6; *CMG* V,4,2, 179, lines 12-17; trans.: R.M. Green, *A translation of Galen's Hygiene*, with an introduction by H.E. Sigerist (Springfield (Ill.) 1951) 250.

68 Galen, III 334 K.; ed. Helmreich, I, 245; trans.. May, 241. For other additions see *Aphorisms* X, 50; XII, 35,40.

69 Other examples of *Aphorisms* containing additional elements are: VI, 94; VIII, 68; IX, 57; X, 50; XII, 35,40; XIII, 2,3,6; XVI, 3; XVII, 5.

70 See *On Asthma*, introduction.

71 Cf. Galen, *De methodo medendi* VIII, 5: X 573 K.

because otherwise their strength would be lost.' This text shows that Maimonides at least occasionally prepared his own salves and medicaments. As we know from the Genizah even famous physicians did this. And in *Aphorisms* XIII, 2,3 when Galen remarks that 'compound purgatives are bad when one of its ingredients has a purgative effect as soon as the purgative enters the body, while another ingredient only has this effect long after its ingestion', Maimonides adds that 'latex plants and scammony purge as soon as they arrive in the body, whereas purgative resins only purge after a <prolonged> period', specifying these resins as opopanax, galbanum, sagapenum, asafetida and gum-ammoniac. At *Aphorisms* XIII, 6 Maimonides comments on Galen's recommendation to rinse the stomach after taking a purgative with barley gruel or groats, that 'the consumption of barley groats once the effect has worn off is something unusual in all the countries he has passed through,' and that 'it induces vomiting because the stomach quickly needs to throw up <even> after the <effect of the> purgative <has worn off>.' In one case Maimonides' personal comment upon a Galenic text gives us additional valuable information about his own training as a doctor. In *Aphorisms* VIII, 68 Galen remarks that the occurrence of diabetes is very rare and that he personally has only seen it twice. Maimonides comments upon this text:

> I too have not seen it in the Maghreb nor did any one of the elders under whom I studied inform me that he had seen it. However, here in Egypt I have seen more than twenty people affected by this disease in approximately ten years. This is to show you that this disease occurs mostly in hot countries. Perhaps the water of the Nile, because of its sweetness, plays a role in this.

This report provides further evidence for the supposition that Maimonides got some sort of formal medical training while residing in the Maghreb.[72] Maimonides' personal note is similar to that featuring in the *Treatise on Asthma* where he remarks, in the context of a notorious medical incident which occurred in the Maghreb, that he studied under a learned physician.[73] Moreover, in the same book he refers to his contacts with physicians from the Maghreb.[74]

Sometimes Maimonides adds elements to the text of Galen in order to explain it. This is, for instance, the case when in *Aphorism* I, 33 he quotes the following text from Galen's *De motu musculorum* II, 5:

> Do not be surprised that most people when asleep perform most voluntary activities such as speaking, screaming, walking and turning from side to side. For those who are awake perform their activities in a state of absent-mindedness so that one may be on one's way with the intention to reach a certain place while one is absent-minded and does not know where one goes until one has completed one's journey.[75]

72 The medieval term 'maghrib' (Islamic West) can refer both to the modern Maghreb (Northwest Africa) and to al-Andalus (Muslim Spain); cf. Blau, 'At our place in al-Andalus, At our place in the Maghreb', *Perspectives on Maimonides. Philosophical and historical studies*, ed. by J. L. Kraemer (Oxford 1991) 293-95; Gerrit Bos, Maimonides, *Treatise on Asthma* IV, 4.

73 *Treatise on Asthma* XIII, 33.

74 See *Treatise on Asthma*, XII, 9; XII, 10.

75 *De motu musculorum* II, 5: IV 440-41 K.

In the introduction to a lengthy exposition of this text Maimonides remarks that these voluntary activities which are performed by one who is asleep and by one who is absent-minded have been extensively verified by Galen in this treatise by his description of his observations but without him giving a reason for it. Maimonides adds that neither did Galen resolve the doubt[76] raised by the question: How can the will of someone who is asleep or absent-minded be abolished and yet he carries out voluntary movements? The critical tone implicit in Maimonides' words becomes more explicit when Maimonides deals with Galen's discussion of the causes of syncope in *Aphorisms* VII, 8. Maimonides criticises him for not being consistent and methodical in his classification of these causes. Maimonides' criticism is aimed at Galen's discussion in *De methodo medendi*. For first Galen mentions a variety of causes for syncope (X 844 K) and then remarks that the remaining causes are four (X 850 K.). Maimonides exclaims: 'I wish I knew why only some causes were given a specific number but not all of them.' He adds that what is even worse is that Galen following these four <causes> remarks: 'And if you want, you can add another one to these four, namely, a bad temperament of the four organs.' Maimonides is once again implicitly critical of Galen when he concludes that since syncope is so serious and is often followed by death, a physician should have a comprehensive knowledge of all its causes and should mention all of them. For if he has a detailed and exact knowledge of its causes, he knows how to treat it when it occurs. And in the name of Abū Marwān b. Zuhr he adds that when a patient suffers from syncope without the physician knowing about it until it <actually> happens and without him warning against it, then this physician is undoubtedly responsible for the death of the patient. The matter is of such vital importance in Maimonides' eyes that he resorts to a summary rearrangement and systematisation of Galen's whole discussion in *Aphorisms* VII, 10:

> Syncope is a quick collapse[77] of the faculties of the body. The existence and permanence of the faculties in their essential nature is dependant on the balance of the pneumata, humours and organs in their quantity and quality. The cause for the collapse of the faculties, namely, syncope can be a change in the quantity or quality of the pneumata or a change in the quantity or quality of the humours or a change in the quantity or quality of the organs.[78] The causes of syncope can thus be limited to three classes and every class has two sub-classes, namely, change in quantity and change in quality. Thus, we have six sub-classes.

On other occasions Maimonides criticises Galen for making contradictory statements. An example is at *Aphorisms* VII, 44 where he quotes the following text from Galen's *De*

76 The Arabic expression 'ḥalla al-shakk' is reminiscent of the title 'ḥall shukūk al-Rāzī'(to solve the doubts raised by al-Rāzī). It refers to a number of works composed by several Islamic physicians, as, for instance, 'Alī ibn Riḍwān (d. 1068), Abū l-'Alā ibn Zuhr (d. 1130-31) and 'Abd al-Laṭīf al-Baghdādī (d. 1231-1232) who came to the defence of Galen against the attack by al-Rāzī (865-932) in his *K. al-shukūk*. See J. Christoph Bürgel, 'Averroes "contra Galenum", Das Kapitel von der Atmung im Colliget des Averroes als ein Zeugnis mittelalterlich-islamischer Kritik an Galen, eingeleitet, arabisch herausgegeben und übersetzt', *NAWG* 1967, 278-290 (285).

77 'a quick collapse': (*suqūṭ... bi-ḥadda wa-sur'a*): *De methodo medendi* XII, 5; X 837 K. The Arabic *bi-ḥadda* is a literal translation of the Greek ὀξεῖα,which has the basic meaning of 'sharp' but can also mean 'quick'.

78 Maimonides' discussion is basically a summary of Galen's lengthy exposition in *De methodo medendi* XII, 5; X 837-45 K.).

tremore, <palpitatione, rigore et convulsione>:[79] 'If the cold <only> moves the outer surface of the body with a shivering movement and shakes it at the moment of the onset of a fever attack but does not move the whole body, this is called 'shivering'. Shivering is an affliction only occurring to the skin. Its significance for the skin is the same as that of rigor for the whole body'. He then comments that: 'In *De febribus* II he said that shivering is something less severe than rigor but more cold.'[80] Even more explicit is his criticism in the following case. Quoting, in *Aphorisms* XV, 3 and 4, Galen's statements that cauterization should only be applied to those bodily parts that do not have depth, i.e., hands, feet and loins,[81] and that in the case of lung ulcers the chest should be cauterized when one despairs of cleansing the lungs through expectoration, Maimonides comments in *Aphorisms* XV, 5 that these two statements would have been mutually exclusive, if Galen had not explicitly stipulated under which exceptional circumstances cauterization to the chest is allowed. However, when on another occasion Galen recommends a quick cauterization to the chest in the case of dropsical patients suffering from lung ulcers, without any further stipulation that might classify this case as exceptional – a text quoted by Maimonides without any further comment in *Aphorisms* XV, 7 – Maimonides attacks him in *Aphorisms* XXV, 18 for being illogical:

> In his commentary to *De aere et aquis* II he says: 'One should not cauterize a part of the body except the hands, feet and loins.' But in Book VII of his commentary on *Epidemics* VI, he says: 'Those who suffer from lung ulcers should be quickly cauterized on their chests.' These are his very words and they annul his previous statement. Consider this.

The critical tone, still subdued and sporadic in these sections of the *Medical Aphorisms,* becomes explicit in the twenty-fifth treatise, in which Maimonides criticises Galen in medical and philosophical issues and charges him with more than forty contradictions in his works and with ignorance in philosophical matters.[82]

It is clear that literal quotations, abbreviations, additions and critical notes play a central role in Maimonides' dealing with the Galenic text. Maimonides abbreviates the repetitious and verbose Galenic text by omitting sections which he considers inessential and thus likely to obscure its real meaning and intention. In turn he adds elements which he believes essential for a correct understanding of the text. Such additions can take the form of explanations, stipulations, practical advice based on Maimonides' personal experience as a physician and evaluation of the text in terms of its medical correctness and incorrectness. But an explicit critique can only be found on rare occasions in the major part of the *Medical Aphorisms.* However, it is the central theme of the twenty-fifth treatise which forms a systematic attack on Galen's inconsistencies. These adaptations of the Galenic material make it very difficult to retrace and identify the original text, especially when the original Greek text has been lost, as in the case of *De humoribus.* Nevertheless, *Medical Aphorisms* is a rich and hitherto virtually unexplored source for the reconstruction of Galenic and pseudo-Galenic texts which no longer exist in the original.

79 *De tremore, palpitatione, rigore et convulsione* 6; VII 612 K.

80 *De diff. febrium* II, 7; VII 357-58 K.

81 Cf. *In Hippocratis De aere aquis locis commentarius* II.

82 For an extensive discussion of the criticism directed against Galen in this treatise see Gerrit Bos, 'Maimonides' Medical Aphorisms: Towards a critical edition and revised English translation', *Korot* 12 (1996-1997) 35-79.

Maimonides' adaptation of Galen's medical works reflects a lifetime of study and practice. It presupposes an intimate familiarity with the texts so that one can retrieve all the relevant material concerning a certain topic sometimes scattered throughout the Galenic corpus. It is an achievement most of us can only perform – and even then with considerable difficulty – thanks to our access to all kinds of indices and especially the *Thesaurus Linguae Graecae*. That Maimonides set great store by the theoretical study of medicine is clear from his statement in the letter to his student Joseph Ben Judah Ibn Shim'on:

> When I come <home> to Fustat, the most that I can do during what is left of the day and the night is to study that which I may need to know from the medical books. For you know how long and difficult this art is for someone who is conscientious and fastidious, and who does not wish to say anything without first knowing its proof, its source <in the literature>, and the type of reasoning (*qiyās*) involved.[83]

Cologne University

83 See D. H. Baneth, *Iggerot ha-Rambam*. Fasc. 1 (Jerusalem 1946) 69-70; Y. Kafih, *Iggerot ha-Rambam* (Jerusalem 1972) 134-35; English translation from the Arabic: Langermann, *Maimonides on the synochous fever*, above, note 64, 176; see also Gerrit Bos, Maimonides, *Treatise on Asthma*, introduction.

THE LOST LATIN GALEN

MICHAEL McVAUGH

My title promises more than it can possibly deliver. 'The lost Latin Galen' would cover an enormous range, after all. When a medieval Latin author of the thirteenth to the fifteenth century introduces a passage with the bare phrase 'ut dicit Galienus,' that unidentified quotation can effectively be said to be 'lost,' though the Galenic original might be findable if we knew where to look. When a Latin author refers to a passage from a lost Galenic title, like *De demonstratione*, we are certainly entitled at least to wonder, wistfully, whether a Latin version once existed and has now disappeared. And, of course, sometimes an author will make explicit reference to a Latin translation of a Galenic work, a translation which nevertheless we cannot identify with any Latin text surviving today. In all these cases we can presume that there was some kind of original Galenic source that is now hidden from us – and that means, of course, that we are referring to a Galen that has been 'lost' only in a weak sense, because in each instance we have a hint of its existence. The complete 'lost Latin Galen' would have to include unattributed passages that we no longer suspect originated in his writings, or translations that were once made and then disappeared without a trace – this Galen is almost inescapably hidden from us and is 'lost' in what you might call the strong sense.

It is not entirely fanciful to imagine that Latin translations of Galenic works could have been made and then been lost in this strong sense, leaving no record whatsoever. An example of this emerges from the *Conciliator* of Pietro d'Abano, a work (completed about 1310) that attempts to mediate a large number of medico-philosophical controversies of the day. Here Pietro refers to, and indeed quotes from, his own translations from Greek into Latin of ten or so Galenic works: *De utilitate particularium*; *De regimine sanitatis*; *De tabe*; *Liber creticorum*; *De anatomia*; *Liber prognosticorum*; *De optima compositione*; *De exercitio cum sphera parva*; and so on. Most of these translations no longer appear to exist in manuscript copies. To be sure, Pietro's translation of *De colera nigra* is well known; Richard Durling identified three copies of what may be Pietro's translation of *De optima compositione*; and Durling also found what may be Pietro's translations of *De bono corporis habitu* and *De exercitio*.[1] However, the rest of the translations that Pietro referred to are apparently lost. We only know about these Latin translations because Pietro happened to mention them in the *Conciliator*, and it is perfectly possible that he made translations of still other Galenic texts, even more deeply lost, which he did not have reason to refer to there.

All this is to explain why this paper cannot presume to be comprehensive. I will touch on many of these kinds of loss in passing, but I will focus on a particular case, the discovery and

1 Richard Durling, 'Corrigenda and Addenda to Diels' Galenica. I', *Traditio* 23 (1967), 468; idem, 'Corrigenda and Addenda to Diels' Galenica. II', *Traditio* 37 (1981), 374, 378.

identification of one particular lost translation, and in the process I will suggest some of the problems we confront in tracking down references to Galen's works in medieval medical authors. The translation in question was of Galen's *Methodus medendi*, and as background something must first be said about the history of that work and its transmission to medieval Europe.

I

The *Methodus medendi* (the *Therapeutic method* or *Method of healing*) has a claim to being the most important of Galen's many works – not simply because of its great length, more than a thousand pages in Kühn's edition of 1825, but for its scope (quoting Vivian Nutton) as 'the most sustained account of Galen's attitude towards medical theory and practice, embracing not only a whole range of varied diseases but also the philosophical arguments and presuppositions that in Galen's view should govern the doctor's therapeutic activities.'[2] To medieval medical practitioners it was further memorable because it presented a number of case histories, vignettes of Galen's encounters with individual patients; to medieval surgeons it was especially significant because its earlier books had so much to say about Galen's surgical practice.

The way in which the *Methodus medendi* was composed by Galen is curious: it was written in two bursts twenty years apart, a gap that had its consequences for the work's content. Nutton has concluded that books I-VI, containing so much of the philosophical polemic and the surgical material, were probably written in the mid-170s; the book was set aside when the friend to whom the work was originally dedicated died, and only in the late 190s did Galen come back to it and complete books VII-XIV (which make much greater use of the case history as an aid to teaching).[3] Nutton has also suggested cautiously that the two halves of the work can be used to understand the progress of Galen's own career: in the second half, twenty years later, he seems more confident of his position, more intolerant of his wealthy patients' vagaries – more at home in Rome, perhaps.[4]

Probably not coincidentally, these two distinct parts of the work – the first six and the last eight books, which from now on I will call Part I and Part II – had different histories as they passed from Greek through Syriac into Arabic, as we learn from Hunain ibn Ishaq in the ninth century. A Syriac translation had been made in the sixth century that Hunain found very unsatisfactory: Part I was, he thought, virtually unusable in that translation, but Part II he tried to touch up by referring to a copy of the Greek text before deciding that he really might as well make a new translation. Unfortunately the only copy of this second translation of Part II was burned shortly afterwards; some years later, therefore, Hunain set about preparing a full Syriac translation of both Parts I and II. For Part I he could find only a single corrupt manuscript in Greek to work from, and he believed that this scarcity of Greek copies was due to the fact that (perhaps because of its focus on surgical matters) this portion of the work had not been an important element in the medical curriculum in the schools of late Alexandria.

2 Vivian Nutton, 'Style and Context in the *Method of Healing*', in Fridolf Kudlien and Richard J. Durling, eds, *Galen's Method of Healing* (Leiden 1991) 1.

3 Ibid., 2-4.

4 Ibid., 21-25.

For Part II, on the other hand, he had several Greek manuscripts, and his Syriac translation was based on their best reading. It was from this new Syriac text that, at the end of the ninth century, Hunain's colleagues at Baghdad finally prepared an Arabic translation; Hunain himself checked over the finished version of Part II. In contrast to the history of the Greek text, Part I was perhaps more popular in Islam than Part II, at least if we can judge from the survival of manuscript material.[5]

Parts I and II of the *Methodus Medendi* also had different histories as they passed on into Latin. The first Latin version of the work was made from the Arabic by Constantine the African in Southern Italy in the late eleventh century; this text, which is known by the name of *Megategni*, is a sharply abbreviated paraphrase of both Parts I and II of the original. Twelfth-century scholars were unhappy with Constantine's failure to translate word for word, as they saw it, and some felt that his work needed to be redone. Two more translations of the *Methodus medendi* were made during that century: Gerard of Cremona, in Toledo, translated the entire work from Arabic under the title *De ingenio sanitatis*, and Burgundio of Pisa, in Italy, translated *only* Part II from Greek under the title *Terapeutica* (from the Greek title, 'therapeutikes methodos'); apparently Burgundio, like Hunain before him, had difficulty finding a Greek text of Part I to work from. Burgundio also left the end of Part II untranslated, stopping in the middle of book XIV, and a translation from Greek of the conclusion was first prepared at the beginning of the fourteenth century by Pietro d'Abano – although, of course, it could always have been read in Gerard's translation from Arabic.[6] There may have been a sense among some early medical readers that the Greek translation of Part II was to be preferred, because many manuscripts of the work survive that contain Part I in Gerard's translation but have suppressed his Part II in favor of Burgundio's version (with its completion by Pietro) – perhaps just as many as survive of Gerard's complete translation of Parts I and II.[7]

II

A few years ago I prepared an edition of Guy de Chauliac's *Inventarium*, a surgical encyclopedia completed in 1363. As part of my project, I made a point of identifying the source of as many as possible of the passages he quoted. This was far from easy. Guy had conceived of his work as a compilation of the existing medical and surgical literature of his day, and he prided himself on the number of authorities he had been able to consult. Unlike most medieval authors, he happily identified all (or almost all) of his references – there are more than three thousand of them. And beyond question, his favorite authority was Galen.

5 On this see Ursula Weisser, 'Zur Rezeption der *Methodus medendi* im *Continens* des Rhazes', in Kudlien and Durling, *Galen's Method*, 123-29; G. Bergsträsser, *Hunain ibn Ishaq über die syrischen und arabischen Galen-Übersetzungen* (Abhandlungen für die Kunde des Morgenlandes, XVII, no. 2) (Leipzig 1925), #20, pp. 14-15; Fuat Sezgin, *Geschichte des arabischen Schrifttums*, vol. 3 (Leiden 1970), 98.

6 According to Pearl Kibre, 'A List of Latin Manuscripts Containing Medieval Versions of the *Methodus Medendi*', revised by R. J. Durling, in Kudlien and Durling, *Galen's Method*, 121, Pietro's contribution begins with the chapter that opens 'Duplex autem est et hoc genus unum quidem . . . ,' which is chapter 12 of the 19 chapters into which the text is divided in Galen, *Opera*, 2 vols, (Venice,1490), 2.219ra.

7 Thus Durling ('Corrigenda I', 474-75) found equal numbers of the two versions among Vatican manuscripts of the work – the composite translation is also that printed in the Venice, 1490, edition of Galen's collected works referred to in n. 6.

But Guy's references did not fully solve the problem of identification for me, for he often neglected to identify exactly where Galen had said something. How, then, can one locate an unattributed fragment, a vague reference to 'dicit Galienus' in Kühn's Galen, by Guy or by any other medieval author? Some concordances to Galen were prepared in the Middle Ages, but identifying a passage in them is a very chancy business indeed. What proved most fruitful was instead to try to guess from Guy's Latin phraseology one or two Greek words that were likely to have been in the original passage (hoping that they had not been disguised or deformed by transmission through Syriac and Arabic) and to look for them in the electronic file (on CD-ROM) of the texts in the *Thesaurus Linguae Graecae*. With patience, I was able in the end to trace in this way virtually every one of Guy's references to 'Galen' back to one or more plausible passages in the original Greek. Then, to confirm these candidates or to choose from among a number of possibilities, I searched them out in the medieval Latin translations of Galen – many of which are available in the great 1490 edition of his works, a few of which have been given modern editions in the *Galenus Latinus* or elsewhere, but several of which had to be consulted in manuscript copies. In this way I was usually able to identify, eventually, not only the works from which Guy was quoting, but (in cases where more than one translation of a Galenic work existed) which translation Guy had chosen to use. This can be a very tedious procedure, but it does work, and it can help us recover some of the 'lost' Galen.

Now in fact Guy took conscious satisfaction from the fact that he had had access to works of Galen that virtually no Latin physician before him had been able to consult, and that in the case of some Galenic works he had been able to read them in recent translations made directly from Greek rather than from Arabic. He was among the earliest medieval medical writers to express an explicit and deliberate preference for translations from Greek. The medical schools of Bologna and Montpellier at the beginning of the fourteenth century had begun to take a new interest in Galen's own writings, but with one exception these were known in translation from *either* Arabic *or* Greek: only *De interioribus* then existed in both an Arabic-Latin and a Greek-Latin version. In the 1290s these versions of *De interioribus* (that is, *De locis affectis*) were of enough interest to Taddeo Alderotti in Bologna that he compared the two systematically, making marginal notes to indicate which translation seemed preferable – but Taddeo's project was unique for his time.[8]

What had altered the situation by Guy de Chauliac's day was the astonishing productivity of a translator in Southern Italy, Niccolò da Reggio. Niccolò's achievement was made possible by the patronage of the Angevin king of Naples, Robert I, for the Angevin library was rich in Greek manuscripts – particularly, it would seem, ones on medicine; this was a subject that had been of special interest to Charles I and was equally so to Robert, who obsessively sent agents out into Southern Italy to search for Greek manuscripts to enlarge his collection. These rulers also commissioned a number of individuals to make Latin translations of the Greek texts in their collection, but the only one from whom an appreciable number of translations has survived is Niccolò da Reggio, who specialized in translating the works of Galen. The catalogue drawn up by Lynn Thorndike in 1946 listed fifty-six Galenic works that Niccolò rendered into Latin; not all his translations are dated, but those that are,

8 Nancy G. Siraisi, *Taddeo Alderotti and His Pupils: Two Generations of Italian Medical Learning* (Princeton 1981) 101-02.

fall between the years 1308 and 1345.[9] The exceptional richness of the royal library where he worked can be judged from the fact that eight of the works Niccolò translated from Greek are today no longer available in that language.[10] Much of the Angevin library was seized in the Hungarian invasion of 1347-48, and a large portion of that (perhaps including those very Greek manuscripts) was lost by shipwreck in the Adriatic on the way back to Hungary. It is not surprising that most later medieval Latin readers remained ignorant of the vast majority of the works he translated.

The fact that no works from the other Angevin translators are known to survive suggested to Robert Weiss that 'perhaps only one transcript was made of each version, namely the one deposited in the Royal Library, and that these unique copies perished with that priceless collection.'[11] Whether or not that is so, Niccolò's translations became known to the West because he sent a copy of at least a portion of his oeuvre to the papal library at Avignon, where Guy de Chauliac was able to consult it.[12] But there are signs that even Guy did not have access to Niccolò's complete output. For example, he refers to one Galenic work as *De motibus liquidis*, the title of its Arabic-Latin translation, not as *De motibus dubiis*, the title given it in Niccolò's retranslation.[13] Again, of Niccolò's fifty-odd translations, Guy refers to only fifteen or so,[14] including (among those that Niccolò dated) *De usu partium* (translated in 1317) and the *Miamir* (also known as *De passionibus uniuscuiusque particule*, translated in 1335), but not others dated 1341 and 1345. So it is conceivable that when Guy came to Avignon in the mid-1340s he found there a collection of Niccolò's earlier translations that had been compiled and given to the papal court in the late 1330s. In any case, Francesco LoParco's suggestion that Niccolò could have presented a copy of his translations to Pope John XXII in 1322 when he came to Avignon in King Robert's entourage is not enough to explain all Guy's quotations.[15] For the moment, let us conclude simply that Niccolò tried to give some circulation to his translations but that, despite his efforts, they did not circulate widely; Guy de Chauliac is one of the very few fourteenth-century medical authors who can be shown to have made use of them.

Guy valued Niccolò's works because, as he declared, they were 'alcioris et perfeccioris stili . . . quam translati de arabica lingua.'[16] He could make such a comparison because Niccolò had retranslated from Greek at least two Galenic works that were already available in translations from Arabic: Galen's commentary on Hippocrates' *Aphorisms*, where Niccolò's completion of a Greek-Latin version begun by Burgundio of Pisa could be compared with the Arabic-Latin translation apparently drawn up by Constantine the African; and Galen's *De*

9 Lynn Thorndike, 'Translations of Works of Galen from the Greek by Niccolò da Reggio (c. 1308-1345)', *Byzantina Metabyzantina* 1 (1946) 213-35.

10 Vivian Nutton, *John Caius and the Manuscripts of Galen* (Cambridge 1987), 30 n. 7.

11 Robert Weiss, 'The Translators from the Greek of the Angevin Court of Naples', *Rinascimento* 1 (1950) 215.

12 Guy de Chauliac, *Inventarium sive Chirurgia Magna*, ed. Michael R. McVaugh, 2 vols, vol. 1 (Leiden 1997) 7.

13 Suggested by Margaret S. Ogden, 'The Galenic Works Cited in Guy de Chauliac's *Chirurgia Magna*', *Journal of the History of Medicine* 28 (1973) 31 n. 20.

14 The complete list is given in ibid., 28-33.

15 Francesco LoParco, 'Niccolò da Reggio antesignano del Risorgimento dell' antichità elleniche nel secolo XIV', *Atti della Reale Accademia di archeologia, lettere e belle arti di Napoli*, n.s. 2 (1910) 262.

16 Guy de Chauliac, *Inventarium* (ed. McVaugh), capitulum singulare, p. 7.

simplicibus medicinis, where Niccolò had translated books I-XI from Greek after Gerard of Cremona had prepared a Latin text of books I-V from Arabic.[17] Thus, for example, at one point Guy quotes *Aphorisms* VI.46 as translated from Greek to the effect that 'anyone who becomes humpbacked *ante pubescentiam* from a cough or asthma will die' – and comments that here the newer translation corrected the meaning of the older version, which had said 'ante iuventutem.'[18] Only twice does Guy specify that he is using the older Arabic-Latin version of the *Aphorisms* in preference to Niccolò's newer one, both times with the implication that Niccolò's very fidelity to Greek syntax had left the passage unclear.[19] And when he quotes the first five books of *De simplicibus medicinis*, whose older translation was less solidly entrenched, he routinely quotes them in Niccolò's version.[20]

III

This returns us, rather circuitously, to Galen's *Methodus medendi*. As we saw, Part I of that work (books I-VI) exists only in the Arabic-Latin version of Gerard of Cremona, called *De ingenio sanitatis*; but Part II (books VII-XIV) exists in both that version from Arabic and the Greek-Latin version of Burgundio of Pisa, called *Terapeutica*. Did Guy de Chauliac show the same preference for a Greek-Latin translation when he quoted *this* work? As you might expect, he did indeed: in fact, at one point he used his access to the newer version to correct his predecessors on a matter of medical detail. Surgeons before him had claimed, on Galen's authority, that wounded patients could be allowed to drink wine from the beginning of their recovery. Guy denied this, saying: 'What misled them was that they used the Arabic translation of the *Methodus medendi*, which reads: 'you only need to avoid giving wine as long as the lesion is hot,' for what it really should say is 'as long as a lesion exists,' as the translation from the Greek makes clear.'[21] Guy might almost be taken as prefiguring Renaissance humanism – except, of course, that he did not know Greek himself and had to take the merits of the Greek-Latin version on faith.

Guy's preference for the Greek-Latin *Methodus medendi*, then, is not surprising. What *is* surprising, indeed astonishing, is that the Greek-Latin version from which he says he is quoting simply does not exist. The lines that he is repeating verbatim occur in book IV of the *Methodus medendi*, which as part of Part I is supposed never to have been translated directly from Greek. Indeed, I have now pursued all Guy's references to Part I, one after another: some of them are paraphrases or summaries that retain only a portion of Galen's original language, some of them are explicitly presented as direct quotations, but none of these passages – 188 altogether![22] – corresponds to the text as translated by Gerard of Cremona. Guy evidently knew and used a translation of Part I of the *Methodus medendi* that has left no other trace in the historical record, and that has utterly vanished; until now it has been lost in the 'strong' sense of the word.

17 Durling, 'Corrigenda I',475, 471; Ogden, 'Galenic Works', 32 n. 23.

18 Guy de Chauliac, *Inventarium* VI.2.3 (ed. McVaugh), 364.

19 Ogden, 'Galenic Works', 26.

20 In *Inventarium* VII.1.5 Guy explicitly corrects the reading of the Arabic version by that of the Greek (ed. McVaugh, 431, lines 11-16).

21 *Inventarium*, IV.7 (ed. McVaugh, 148, lines 36-42); Ogden, 'Galenic Works', 27.

22 Guy makes 5 references to book I, 2 to book II, 32 to book III, 48 to book IV, 36 to book V, and 65 to book VI.

Where might Guy have come upon this translation? Who could have composed it, and why has it disappeared? The second of these questions, at least, seems to have an obvious answer, that this is one of the lost translations from the Greek that Pietro d'Abano claimed to have prepared. We have already encountered Pietro as a self-proclaimed translator of Galen, the man who completed Burgundio's Greek-Latin translation of book XIV of the *Methodus medendi*, but there is a case to be made for his having done much more than this, depending on how we interpret his reference in the *Conciliator* to a passage 'in principio . . . terapeutices artis [that is, the *Methodus medendi*] *sicut transtuli*.'[23] Lynn Thorndike supposed from this that Pietro not only finished book XIV for Burgundio but was here revealing that he had also prepared a Greek-Latin version of books I-VI, thus completing Burgundio's translation at the beginning as well as at the end.[24] And since the passage that Pietro proceeds to quote seems to correspond very closely to a passage that Guy happens to quote from book I, and since both are very different from the standard Arabic-Latin translation of the same passage, it might seem highly probable that our lost translation of the *Methodus medendi* had been composed by Pietro d'Abano, sometime before 1310.[25]

But in fact things are not this simple. As it happens, Pietro quotes the *Methodus medendi* not just this once in the *Conciliator* but something like forty times. He refers to the work under three titles: 'De ingenio sanitatis,' 'Terapeutice ars,' and 'Curative ars.' And whatever the title, with the *one exception* of the passage I have just referred to, every time he quotes the work he is demonstrably referring to the composite version that circulated widely in his day, comprising Part I in the Arabic-Latin translation and Part II in the Greek-Latin translation. The version that always leapt to his pen, the definitive version to his mind, was – with that one exception – the traditional version of Gerard and Burgundio.

How can we explain this? I believe that Pietro's reference to his 'translation' refers not to a translation of the complete work but to one of this passage only – that at this particular point, he chose to consult a Greek manuscript to which he had access and to translate these few words into Latin because he was dissatisfied with the Arabic-Latin version. If we look at his translation of this phrase and compare it closely with that given in Guy's encyclopedia, we find small but significant differences: Pietro translates 'δέονται' as 'indigent,' Guy as 'egerent,' for example.[26] I suggest that the resemblance between the two passages is merely coincidental, a product of the simple and relatively unequivocal phraseology of the Greek being quoted, and

23 On the date, see Lynn Thorndike, 'Translations of Works of Galen from the Greek by Peter of Abano', *Isis* 33 (1942) 649 n. 1. Following Thorndike (650 n. 13), I have quoted Pietro as saying 'transtuli' ('I translated'), although the printed editions all say 'transtulit' ('he translated'). Thorndike goes on (p. 651) to suggest that the *Therapeutica ars* is the same work as the *De regimine sanitatis* to which Pietro refers at least twice – 'De regimine sanitatis de verbo ad verbum'; 'de regimine sanitatis primo' – but *De regimine sanitatis* is the name by which *De sanitate tuenda* was already known, and Pietro's references are actually to that work (see the following paragraph).

24 Thorndike, 'Translations by Pietro', 652.

25 This was suggested by Ogden, 'Galenic Works',32-33 n. 25. For the comparative texts, see below, n. 26.

26 Pietro: 'Si neque geometria neque astronomia aut dialectica neque alia aliqua doctrina bonarum medici fieri debentes indigent . . .' (*Conciliator*, Diff. 1 (Venice, 1565; rpt. Padua, 1985), f. 3va). Guy: 'Si enim neque geometria neque astronomia neque dyalectica neque aliqua alia doctrina bonorum egerent medici . . .' (*Inventarium*, ed. McVaugh, 9). *De ingenio*: 'si . . . dicens in medicina discenda nihil dyalectice vel geometrice astronomie et ceterarum liberalium artium esse necessarium' (*Opera Galieni* [above, n. 6], 2.167ra). Galen: 'εἰ γὰρ οὔτε γεωμετρίας οὔτε ἀστρονομίας οὔτε διαλεκτικῆς οὔτε μουσικῆς οὔτε ἄλλου τινὸς μαθήματος τῶν καλῶν οἱ μέλλοντες ἰατροὶ γενήσεσθαι δέονται' (Carolus Gottlob Kühn, *Claudii Galeni Omnia Opera* (Leipzig, 1821-33; hereafter K), 10.5).

that there is no good evidence that Pietro either prepared or used the translation of the *Methodus medendi* from which Guy was quoting so extensively fifty years later.

If this is correct, it may also help explain why it has been so hard for scholars to identify manuscripts of the other Galenic works Pietro says he has 'translated': perhaps in *all* these instances he is referring simply to passages he has extracted and translated ad hoc, not to the whole work. Indeed, by comparing Pietro's language with that of previous translations of these works we can get some sense of how he may have worked. In *Conciliator* diff. 18, for example, he quotes Galen as saying in the first book of what he calls '*De regimine sanitatis*' that 'Simplex sanitas et perfecta quae certissimam habet eucrasiam vel nunquam fuit in animalis corpore,' and this proves to be a direct quotation from the already extant Greek-Latin translation of *De sanitate tuenda* I (corresponding to K 6.29 lines 1-3) . He then goes on to offer his own translation of what he takes to be the same passage, which turns out instead to be the Greek text in K 6.25 lines 10-11. I suggest that Pietro was looking hastily through his Greek copy, searching for something like the words 'zoon somati' to correspond to the 'animalis corpore' of the translation, and in fact those words do appear in both passages. In fact, he was going through a procedure of identification very like the method that can be used today to identify Galenic passages in the electronic file of the *Thesaurus Linguae Graecae*. If I am right about all this, therefore, there is now no hope of ever finding most of the long-lost Galenic translations of Pietro d'Abano – they simply never existed.

IV

I have already suggested that the Greek-Latin translation of Part I of the *Methodus medendi* cited by Guy is another translation by Niccolò da Reggio whose existence has previously been unsuspected, although admittedly the case to be made must be woven together from likely assumptions. We need to assume, first, that Niccolò was able to find a copy of the complete Greek text (or at least of Part I) in King Robert's library – which is not terribly difficult to imagine, given the king's obsession with collecting. Next, we have to assume that the copy Niccolò prepared for the king was lost, like so many other things, in the dispersal of the royal library, but that he had sent a second copy to Avignon, where Guy de Chauliac was able to consult it. It has sometimes been supposed that MS Paris, BN lat. 6865, which was in the possession of a member of the papal court at Avignon in 1353 and which contains more than thirty of Niccolò's translations, might be the manuscript sent by Niccolò and used by Guy – and of course the *Methodus medendi* is not present in this manuscript.[27] However, even if this particular manuscript was actually sent by Niccolò to the court, it cannot be the only copy of his works that he sent there because it does not include all the translations of his that Guy cites: Guy also used Niccolò's translations of *De constitutione artis medicine*, *De facile acquisibilibus*, *De sanitate tuendi*, *De utilitate partium*, *Miamir*, and the commentary on Hippocrates' *Aphorisms* – none of which is in BN lat. 6865. These would therefore have

27 LoParco, 'Niccolò da Reggio', 262-63; Vivian Nutton, ed., *Galen On prognosis* (*CMG* V.8.1.1; Berlin, 1979), 27. Neither comments on the fact that this manuscript is not made up exclusively of translations by Niccolò, or tries to explain why he should have wanted to send other translations besides his own. Thorndike, 'Translations by Niccolò da Reggio', 218 n. 17, identifies the authors of some of the other translations contained in this manuscript. I have discussed this question at greater length in 'Niccolò da Reggio's Translations of Galen and Their Reception', forthcoming.

had to be included in a second codex, now lost – which could also perfectly well have included the first six books of the *Methodus medendi*.

Finally, we need to assume that those six books were little copied by later scribes and hence did not manage to survive, and this is not entirely far-fetched either. By Niccolò's day (the mid-fourteenth century) the language of the older Arabic-Latin versions was so well entrenched in university curricula that masters seem to have preferred not to cope with the newer translations. This is the case, for example, with Galen's commentary on the *Aphorisms*: Pearl Kibre has identified 76 fourteenth- or fifteenth-century copies of its translation by Constantine the African, but only 10 of the translation begun (apparently) by Burgundio of Pisa and completed by Niccolò early in his career, in 1314.[28] Moreover, the prominence given to surgery in Part I of the *Methodus medendi* made it of relatively little importance to many medical faculties in the later Middle Ages, especially as surgery was becoming more and more a nonacademic craft in many parts of Europe. It is still remarkable, but not perhaps astonishing, that – assuming with me that it once did exist – a translation by Niccolò of Part I of the *Methodus medendi* should not have survived.

V

So there is my reconstruction: my fragments reveal a *Methodus medendi* that was among the Galenic works translated from Greek by Niccolò da Reggio and sent to Avignon, but that, perhaps because of its narrowly surgical focus, could not compete successfully with the traditional Arabic-Latin version of Gerard of Cremona. If any copies were made, they were too few to allow it to survive, and it was lost. And yet now of course it is not quite lost, because it has left its imprint in Guy de Chauliac's *Inventarium*. As I discovered more and more citations to Part I in that work, I found myself intrigued by the problem of reconstructing the text he had used, which was in the end relatively straightforward: it was no great task to assemble the citations and to put them in order by comparing them with Kühn's Greek text. Guy incorporated only seven passages from the first two books, which are concerned more with abstract methodology than with the details of practice, but he quoted much more extensively from the later books with their surgical detail: 65 times from book VI, and 48 times from book IV. The citations are often brief, but not infrequently he quotes a sentence or two in full; and, once, a combination of quotations from Guy's text corresponds to nearly two consecutive pages in Kühn.[29] A partial reconstruction of the text of the vanished translation is thus certainly possible.

In the Appendix to this paper (pp. 163-64 below), I have supplied the longest of these Latin fragments, with the corresponding Greek. Some day, perhaps, these and the other passages quoted by Guy may help reconstruct the Greek textual tradition. More interestingly from my perspective, they may eventually make it possible to decide whether my hypothesis about Niccolò da Reggio as the author of the lost translation is correct. In the last twenty years or so scholars have begun the process of analyzing different individuals' styles of translation

28 Pearl Kibre, *Hippocrates Latinus* (New York, 1985), comparing copies listed on pp. 55-61 and 61-62. On the proposed date, see Thorndike, 'Translations by Niccolò da Reggio', 219.

29 The Greek text in K 10.238-40 corresponds to the Latin in *Inventarium* IV.1.1 (ed. McVaugh), 213 lines 34-39 and 212 lines 3-22; see the Appendix below, pp. 163-64.

into Latin, from both Arabic and Greek; but conceivably because there is so much material to analyze, Niccolò da Reggio's techniques of translation have not yet been systematically investigated and characterized.[30] In five years, or ten, or twenty, this will no doubt change, and when that happens someone will perhaps be able to identify the distinctive signs of Niccolò's word choice and sentence structure in these fragments. When that happens – if that happens – *one* part of the lost Latin Galen will no longer be quite so lost. We can only guess how much more, and of what kind, may remain to be discovered.

University of North Carolina, Chapel Hill

30 A start on Niccolò has been made by Irmgard Wille, 'Zur Übersetzungstechnik des Nikolaus von Rhegium in Galens Schrift De temporibus morborum,' *Helikon* 3 (1963) 259-77; and see also the literature cited in Carlos J. Larrain, 'Galen, *De motibus dubiis*: Die lateinische Übersetzung des Niccolò da Reggio', *Traditio* 49 (1994), esp. 171-75. Useful material on William of Moerbeke's technique (Moerbeke is known to have translated one work of Galen, *De alimentis*, which he finished in 1277) can be studied in Proclus, *Commentaire sur le Parménide de Platon, traduction de Guillaume de Moerbeke*, ed. Carlos Steel (Leiden 1982); and in Alexander of Aphrodisias, *De fato ad imperatores, version de Guillaume de Moerbeke*, ed. Pierre Thillet (Paris 1963). For Burgundio, see Richard J. Durling, ed., *Galenus Latinus*: I. Burgundio of Pisa's translation of Galen's Peri kraseon, 'De complexionibus' (Berlin 1975); II. Burgundio of Pisa's translation of Galen's Peri ton peponthoton topon, 'De interioribus' (Berlin 1992).

APPENDIX

A Galen, *Methodus medendi* IV.2 (K 10.238-40)

[ἐπειδὰν δὲ καὶ μέχρι πλέονος ἡ διάθεσις ἐκτείνηται, σκέψις ἐνταῦθα γίνεται πότερα περικοπτέον ἅπαν τὸ παρὰ φύσιν ἐστὶν, ἢ θεραπευτέον ἐν χρόνῳ. καὶ δῆλον ὡς καὶ τῇ τοῦ κάμνοντος εἰς τοῦτο προσχρῆσθαι δεῖ προθυμίᾳ· τινὲς μὲν γὰρ ἐν χρόνῳ πλείονι θεραπεύεσθαι βούλονται χωρὶς τομῆς· ἔνιοι δὲ πᾶν ὁτιοῦν ὑπομένειν εἰσὶν ἕτοιμοι τοῦ θᾶττον ὑγιᾶναι χάριν.] [οὕτω δὲ κἀπὶ τῶν ἐπιρρεόντων τοῖς ἡλκωμένοις μέρεσι μοχθηρῶν χυμῶν ἡ μὲν ὡς ἡλκωμένων ἴασις ἐν τῷδε λελέξεται, ἡ δ' ὡς κακοχυμίας ἢ πλήθους ἐν τοῖς ἰδίοις ἐκείνων λογισμοῖς. ὅταν οὖν ὀλίγῳ τε πλείω καὶ μὴ πολλῷ φαυλότερος ᾖ τοῦ κατὰ φύσιν ὁ ἐπιρρέων τοῖς ἡλκωμένοις χυμὸς, ἀποτρέπειν αὐτὸν καὶ ἀναστέλλειν προσήκει, στύφοντά τε καὶ ψύχοντα τὰ πρὸ τῶν ἡλκωμένων χωρία. χρὴ δὲ καὶ τὴν ἐπίδεσιν ἄρχεσθαι μὲν ἀπὸ τοῦ πεπονθότος, ἐπινέμεσθαι δὲ ἐπὶ τὸ ὑγιὲς, ὡς ἐν τοῖς κατάγμασιν ἐκέλευσεν ὁ Ἱπποκράτης. ἀλλὰ καὶ τὰ τοῖς ἕλκεσιν αὐτοῖς προσαγόμενα φάρμακα ξηραντικώτερα τῶν τοῖς ἁπλοῖς ἕλκεσι προσαγομένων ὑπαρχέτω. μὴ δυναμένης δὲ ὑπὸ φαρμάκων κρατηθῆναι τῆς ἐπιρροῆς, τὴν αἰτίαν αὐτῆς ἐπισκεψάμενον, ἐκείνην ἐκκόπτειν πρότερον. εἰ μὲν οὖν δι' ἀτονίαν τινὰ τοῦ δεχομένου τὸ ῥεῦμα μορίου τοῦτο συμβαίνει, ταύτην ἰατέον· εἴη δ' ἂν ἔτι τοῦτο τῶν ἡλκωμένων μορίων οἰκεία τις ἴασις· εἰ δὲ διὰ πλῆθος ἢ κακοχυμίαν ἤτοι παντὸς τοῦ σώματος ἤ τινος τῶν ὑπερκειμένων μορίων, ἐκεῖνα πρότερον ἐπανορθωτέον. ἡ μὲν οὖν ἀτονία τοῦ μέρους,] δι' ἣν ἐπ' αὐτὸ πλείους τοῦ δέοντος ἀφικνοῦται χυμοὶ, πάντως μὲν ἐπὶ δυσκρασίᾳ γίνεται, οὐ μὴν ἁπάσῃ γε ἕπεται ἀτονία δυσκρασίᾳ·

B Guy de Chauliac, *Inventarium* IV.1.1 (ed. McVaugh, pp. 213, 212)

Si vero ad amplius disposicio extenditur, dicit Galienus in 4 ° Terapentice quod scrutacio debet fieri utrum abscindendum sit totum quod preter naturam vel curandum in tempore (supple cum medicinis acutis). Et palam quoniam et laborantis animo oportet uti: quidam enim in tempore ampliori curari volunt absque incisione, quidam vero quodcumque sustinere prompti sunt, gracia cite sanacionis. Nichilominus ipse dicit inferius: Etenim promptissimum est ad scindendum, maius vero et artificialius est sanare farmacis. ...

Ita dicit Galienus in 4 ° Terapentice: Et influentibus ulceratis particulis pravis humoribus, sanacio quidem que est ut ulceratarum hic dicetur. Quod vero est cachochimie vel plectorie in propriis istarum raciociniis, iam est dicta superius in tractatu apostematum, et specificat modum de utroque. Quando igitur paulo amplior et non multo pravior fuerit eo quod secundum naturam qui influit ulceratis humor, prohibere ipsum et repercutere convenit, stipticando et infrigidando eas que sunt ante ulceratas particulas. Oportet autem et super ligacionem (supple repulsivam) incipere quidem a paciente, superpossidere vero sanam, sicut in fracturis iussit Ypocras: talis enim ligatura constringit meatus per quos materia ad particulas influitur. Ac vero et in ulceribus ipsis afferemus farmaca siccanciora hiis, que simplicibus scilicet vulneribus offeruntur (ecce differenciam); nequeunte vero a farmacis detineri fluxione (ecce bonam practicam), causam eius scrutando illam oportet abscindere prius.

Si igitur propter inbecillitatem aliquam suscipientis reuma particule hoc accidit, hanc sanandum; erit autem ad hoc harum particularum ulceratarum propria quedam sanacio. Si vero propter multitudinem vel cacochimiam tocius corporis vel alicuius superiacencium particularum illa prius corrigendum; inbecillitas quidem particule discrasia fuit. Quomodo autem curatur particula vel universum corpus mittens materiam que influit, in apostematibus superius fuit dictum.

APPENDIX – A SHORT HANDLIST OF TEXTS NOT PRINTED IN K.[1]

Administrationes anatomicae IX-XV
ET M. Simon, Leipzig, 1906
T W. H. L. Duckworth, M. C. Lyons, B. Towers, Cambridge, 1962
T I. Garofalo, Turin, 1991

De anatomia mortuorum
I. Ormos, 'Bemerkungen zur editorischen Bearbeitung der Galenschrift *Über die Sektion toter Lebewesen'*, in *Galen und das hellenistische Erbe*, ed. J. Kollesch, D. Nickel, Stuttgart, 1993, 165-172 (hinting also at the possible authenticity of an unpublished *De anatomia vivorum.*)

De causis continentibus
EC K. Kalbfleisch, Diss., Marburg, 1904
ET M. C. Lyons, *CMG Suppl. Or.* II, Berlin, 1969

De causis procatarcticis
E K. Bardong, *CMG Suppl.* 2, Leipzig, 1937
ECT R. J. Hankinson, Cambridge, 1998

De consuetudinibus
E F. R. Dietz, Leipzig, 1832
E I. von Müller, Leipzig, 1891
E J. Schmutte, F. Pfaff, *CMG Suppl.* III, Leipzig, 1941

De demonstratione (Fr.)
E I. von Müller, *Abh. Akad. Wiss. München*, 1895, 403-78
Cf. G. Strohmaier, 'Bekannte und unbekannte Zitate in den *Zweifeln an Galen* des Rhazes', *Text and Tradition*, ed. K. D. Fischer, D. Nickel, P.Potter, Leiden, 1998, 263-89

[1] E = Edition; T = Translation; C = Commentary. Fr. indicates that the edited portion is incomplete.
This list does not include the several brief quotations in Arabic authors from lost Galenic works, e.g. his views on Christians in his Platonic commentaries, cf. R. Walzer, *Galen on Jews and Christians* (Oxford, 1949); papyrus fragments possibly deriving from lost works; or references within the Galenic Corpus itself to works otherwise lost, e.g. *De medicina apud Homerum*, cf. F. Kudlien, 'Zum Thema Homer und die Medizin', *Rheinisches Museum* 108, 1965, 293-99.

De diaeta in morbis acutis secundum Hippocratem
ET M. C. Lyons, *CMG Suppl. Or.* II, Berlin, 1969

De dolore evitando (Fr.)
E A. S. Halkin, *Proc. Amer. Acad. Jewish Research* 14, 1944, 60-65, 110-115
ET M. Zonta, *Un interprete ebreo della filosofia di Galeno*, Turin, 1995, 113-123

De experientia medica
ET R. Walzer, Oxford, 1944
(repr. M. Frede, *Galen, Three treatises on the nature of science,* Indianapolis, 1985

De moribus (Fr.)
E P. Kraus, *Bull. Fac.Arts Egypt. Univ.* 5, 1937, 1-51
T J. N. Mattock, *Festschrift Richard Walzer*, Oxford, 1972,1-51
T M. Zonta, *Un interprete ebreo della filosofia di Galeno*, Turin, 1995, 29-79, 125-44

De motibus dubiis
EC C. J. Larrain, *Traditio* 49, 1994, 171-233; 51, 1996, 1-41

De nominibus medicis
ET M. Meyerhof, J. Schacht, Abh. Akad. Wiss. Berlin, 1931

De optimo medico cognoscendo
ETC A. Z. Iskandar, *CMG Suppl. Or.* IV, Berlin, 1988

De partibus artis medicativae
E H. Schöne, Greifswald, 1911
ET M. C. Lyons, *CMG Suppl. Or.* II, Berlin, 1969

De partium homoeomerium differentia
ETC G. Strohmaier, *CMG Suppl. Or.* III, Berlin, 1970

De possibilitate (Fr.)
ET N. Rescher, *Alexander against Galen on motion,* Islamabad, 1965, 69-70

De propriis placitis
ETC V. Nutton, *CMG* V 3,2 Berlin, 1999

De septimestri partu
E H. Schöne, *Quellen und Studien Gesch. Naturwiss. Med.* 3, 1932/3, 120-138
ET R. Walzer, *Rivista di Studi orientali* 15, 1933, 323-357;

De victu attenuante
E K. Kalbfleisch, Leipzig, 1898
E K. Kalbfleisch, *CMG* V 4,2, Leipzig, 1923
ET N. Marinone, Turin, 1973
TC W. Frieboes, F. W. Kobert, Breslau, 1903
T P. N. Singer, Oxford, 1997

De visu (Fr.)
S. Vardanian, 'Galen und die mittelalterliche Medizin', in, *Galen und das hellenistische Erbe*, ed. J. Kollesch, D. Nickel, Stuttgart, 1993, 193-204.

De voce (Fr.)
E H. Baumgarten, Diss., Göttingen, 1962

Compendium Timaei Platonis
E P. Kraus, R. Walzer, London, 1951

In Hippocratis De aere aquis locis commentarii
ET A. Wasserstein, Jerusalem, 1982 (Fr.)
ET G. Strohmaier, *CMG Suppl. Or., forthcoming*
T G. Toomer, 'Galen on the astronomers and astrologers', *Archive for the history of exact science* 32, 1985, 193-206 (Fr.)

In Hippocratis De humoribus commentarii (Fr.)
E K. Deichgräber, 'Hippocrates De humoribus in der Geschichte der griechischen Medizin' *Abh. Akad. Wiss. Mainz* 1972, 38-55

In Hippocratis Epidemiarum librum II commentarii
T F. Pfaff, *CMG* V 10,1, Leipzig, 1934

In Hippocratis Epidemiarum librum VI commentarii VI-VIII
T F. Pfaff, *CMG* V 10,2,2, Leipzig, 1940/1956

In Hippocratis Iusiurandum commentarius (Fr.)
T F. Rosenthal, *Bulletin of the History of Medicine* 30, 1963, 52-87

In Hippocratis De octomestri partu commentarius (Fr.)
T H. Grensemann, *Die hippokratische Schrift De octomestri partu,* Diss., Kiel, 1960, 41-53

In Platonem Timaei Commentarius (Fr.)
ETC C. Daremberg, Paris, 1948
EC H. O. Schröder, P. Kahle, *CMG Suppl.* I, Leipzig, 1934
ETC C. J. Larrain, Stuttgart, 1992 (but dubious)

Institutio logica

E M. Mynas, Paris, 1844
E K. Kalbfleisch, Leipzig, 1896
TC J. Mau, Berlin, 1964
TC J. S. Kieffer, Baltimore, 1964
T I. Garofalo, M. Vegetti, *Galeno. Opere Scelte,* Turin, 1978

Quod primum movens immotum (Fr.)

ET N. Rescher, *Alexander against Galen on motion,* Islamabad, 1965

Subfiguratio empirica

E M. Bonnet, Diss., Bonn, 1872
EC K. Deichgräber, *Die griechische Empirikerschule,* Berlin, 1930/1965, 7-19
T M. Frede, *Galen, Three treatises on the nature of science,* Indianapolis, 1985
TC J. Atzpodien, Husum, 1986

Synopsis librorum suorum De methodo medendi,

T I. Garofalo, forthcoming

Synopsis librorum anatomicorum Marini (Fr.)

ECT J. Kollesch, *CMG Suppl.* V, Berlin 1964, 30-33, 68-75

INDEXES

I NAMES AND TOPICS

II GALENIC PASSAGES [1]

[1] Individual passages are listed by book and chapter, and by the volume and page in Kühn's edition. Material not in K. is cited from the editions given on pp. 165-168.